The Privatisation of Biodiversity?

NEW HORIZONS IN ENVIRONMENTAL AND ENERGY LAW

Environmental law – including the pressing considerations of energy law and climate change – is an increasingly important area of legal research and practice. Given the growing interdependence of global society and the significant steps being made towards environmental protection and energy efficiency, there are few people untouched by environmental and energy lawmaking processes.

At the same time, environmental and energy law is at a crossroads. The command and control methodology that evolved in the 1960s and 1970s for air, land and water protection may have reached the limit of its environmental protection achievements. New life needs to be injected into our environmental protection regimes – perhaps through the concept of sustainability in its environmental, economic and social forms. The same goes for energy policy and law, where liberalization, environmental protection and security of supply are at the centre of attention. This important series seeks to press forward the boundaries of environmental and energy law through innovative research into environmental and energy law, doctrine and case law. Adopting a wide interpretation of environmental and energy law, it includes contributions from both leading and emerging international scholars.

Titles in the series include:

Polar Oceans Governance in an Era of Environmental Change
Edited by Tim Stephens and David L. VanderZwaag

The Fragmentation of Global Climate Governance
Consequences and Management of Regime Interactions
Harro van Asselt

Renewable Energy Law in the EU
Legal Perspectives on Bottom-up Approaches
Edited by Marjan Peeters and Thomas Schomerus

Environmental Enforcement Networks
Concepts, Implementation and Effectiveness
Edited by Michael Faure, Peter De Smedt and An Stas

Regulation of the Upstream Petroleum Sector
A Comparative Study of Licensing and Concession Systems
Edited by Tina Hunter

Regional Environmental Law
Transregional Comparative Lessons in Pursuit of Sustainable Development
Edited by Werner Scholtz and Jonathan Verschuuren

Earth Governance
Trusteeship of the Global Commons
Klaus Bosselmann

Waste Policy
International Regulation, Comparative and Contextual Perspectives
Alexander Gillespie

Shale Gas and the Future of Energy
Law and Policy for Sustainability
Edited by John C. Dernbach and James R. May

The Privatisation of Biodiversity?
New Approaches to Conservation Law
Colin T. Reid and Walters Nsoh

The Privatisation of Biodiversity?

New Approaches to Conservation Law

Colin T. Reid

Dundee Law School, University of Dundee, UK

Walters Nsoh

Birmingham Law School, University of Birmingham, UK

NEW HORIZONS IN ENVIRONMENTAL AND ENERGY LAW

Edward Elgar PUBLISHING

Cheltenham, UK • Northampton, MA, USA

Published by
Edward Elgar Publishing Limited
The Lypiatts
15 Lansdown Road
Cheltenham
Glos GL50 2JA
UK

Edward Elgar Publishing, Inc.
William Pratt House
9 Dewey Court
Northampton
Massachusetts 01060
USA

A catalogue record for this book
is available from the British Library

Library of Congress Control Number: 2016935781

This book is available electronically in the **Elgar**online
Law subject collection
DOI 10.4337/9781783474448

FSC
www.fsc.org
MIX
Paper from
responsible sources
FSC® C013056

ISBN 978 1 78347 443 1 (cased)
ISBN 978 1 78347 444 8 (eBook)

Typeset by Columns Design XML Ltd, Reading
Printed and bound in Great Britain by TJ International Ltd, Padstow

Contents

Preface

The project which led to this book was inspired by a number of factors. First, the continuing loss of biodiversity, despite the volume of law and policy designed to prevent this, suggested that it may be time for a reconsideration of how we go about the task of conserving biodiversity for our own sake and that of future generations. Secondly, across many areas of environmental law there have been moves away from the established 'command and control' model of regulation to introduce a more varied range of approaches, including economic instruments and market-based mechanisms. So far there has been only a limited application of these to biodiversity issues and limited exploration of what that might entail. Yet it seems inevitable that more attention will be paid to such possibilities, especially in the dominant climate of deregulation and austerity. This gives rise to the danger that seemingly attractive ideas might be blindly applied without proper appreciation of the special features of biodiversity in comparison to other environmental issues (such as greenhouse gases) or of the risks as different mechanisms move from playing a supplementary to a central role. Thirdly, in recent years there has been a major movement towards discussing biodiversity in terms of ecosystem services and natural capital, which opens the door to conceiving of the value and role of biodiversity in ways quite different from those embedded in current regulatory and legal frameworks. The final spur was the realisation that there was a comparative absence of discussion of such issues from a legal perspective, despite the wealth of literature on the economic, ecological and ethical aspects of 'valuing nature'. Our aim is to make a contribution to filling that gap.

The focus, therefore, is on the legal issues and questions arising in the design and operation of systems, not the underlying economic, regulatory and ethical arguments. Writing this book would have been easier (and reading it possibly more exciting) if we held passionately fixed views on one extreme or the other. A simpler account could be given if we saw an economic approach as, on the one hand, the ideal way of dealing with all the relevant regulatory problems, or on the other hand as requiring a fundamentally misconceived and unacceptable attempt to value nature in financial terms alone. Instead, and along with most of the stakeholders we have come across, from many different backgrounds, our view is more pragmatic. Market-based solutions will not be appropriate for all aspects of biodiversity conservation, but they may offer opportunities for improving on the current position, where despite the laws that are in place, biodiversity too often is on the losing side when competing against other interests. A lot depends on the specific scope and design of particular schemes and it is considering these design challenges which lies at the heart of this book.

We speak of a shift from the more established regulatory style to the 'new' mechanisms and approaches discussed in this book, but that description must be seen simply as a useful short-hand and certainly not as a wholly accurate term. As is made clear in the following pages, around the world there are many schemes currently in operation that apply the methods explored below. The management agreements which play an important part in looking after habitat in the United Kingdom are a form of payment for ecosystem services (based on something akin to a short-term conservation covenant); biodiversity offsetting and bio-banking schemes are established in Australia and the United States; and conservation easements are widely used across the United States.[1] Nevertheless, even in such contexts, the mechanisms are a more recent addition to the more established 'command and control' approach and the contrasts between that approach and the 'new' one are central to the analysis here.

THE BOOK

The book begins with an overview of the factual, legal and regulatory background to biodiversity conservation and then in Chapter 2 offers some consideration of a number of pervasive issues which arise, with differing intensity, across all of the topics discussed in the book. The next five chapters then offer more detailed consideration of various mechanisms through which the law can be used to further conservation, before the underlying ethical issues are explored in Chapter 8. The final chapter offers some reflection on what has gone before. Throughout the book we have generally avoided discussion of the further substantial challenges, but also opportunities, which arise when mechanisms are designed to operate at an international level rather than within single jurisdictions. Likewise, we touch on, but do not fully develop, the particular challenges

[1] See Chapters 3, 4 and 5 respectively.

faced in adopting any conservation measures across the developing world where, for example, the ownership of much land may be unclear, or at least the formal legal interests may not properly reflect the interests of the people who depend on it.

Given our academic histories and the place where we are working, we recognise that we are writing from a background steeped in the thinking of western (especially common law) legal systems and western ethical debates. Different philosophical, cultural and legal traditions in other parts of world may have a lot to offer, but we are in no position to do justice to that contribution. Similarly, we recognise that our default position is to consider issues from a UK perspective. This again has limitations, but does offer a consistency of view in exploring the ramifications of various schemes, rather than jumping between jurisdictional viewpoints when the varying contexts of apparently similar measures can produce quite different results. Since Professor Reid has written about the implications for environmental law of devolution within the United Kingdom, it should be pointed out that strict accuracy has been sacrificed for the sake of an easier flow of the text and there has not been rigour in distinguishing in detail between matters which are ones for UK law, those for the law of Great Britain (i.e. Scotland, England and Wales, without Northern Ireland) and those for the law of each jurisdiction separately.

In carrying out our research we have drawn on literature well beyond legal sources, particularly in ecology, environmental economics and philosophy. It should be stressed, though, that we have not attempted any comprehensive survey of all the relevant works across these fields. We send our apologies to the authors whose work we would have benefited from but have overlooked and omitted.

The book is very much the product of joint work, but in preparing the book the initial draft of Chapter 3 was prepared by Dr Nsoh and the remainder by Professor Reid.

ACKNOWLEDGEMENTS

This book is largely the result of an AHRC-funded research project, The Privatisation of Biodiversity?: New Approaches to Nature Conservation Law (AH/I022884/1AHRC), which allowed Dr Nsoh to be employed at the University of Dundee from April 2012 until April 2014 and without which the research could not have been completed. That award also enabled us to hold a major conference in Dundee in February 2014 and to attend various other conferences, workshops and meetings where we learned so much from the other contributors and participants. The comments of the advisory panel for the project were also very much valued: Mary F.B. Christie, Charles Cowap, Tom Huggon, Dr Maggie Keegan and Professor Andrew Watkinson.

We are grateful to the publishers for their indulgence as the preparation of this book has dragged on longer than anyone had envisaged. Exceeding the original timetable is hardly an unusual feature for academic works, but was something we sought to avoid. Some explanation, if not excuse, is provided by three factors: the fact that the UK Border Agency wrongly refused Dr Nsoh a visa when he started work at Dundee, requiring his employment to be suspended for several weeks until the position was rectified; the commitments of Professor Reid as a member of the Law sub-panel for REF 2014, which imposed a heavy workload just when the writing should have been getting under way as the funded period of the project was coming to an end; and the fact that Dr Nsoh had to cope with the disruption and many other commitments arising from his moves to start a career as a Lecturer in Law at one institution and then in the middle of 2015 to his current post at Birmingham.

Both authors are deeply indebted to the many people from many backgrounds whose comments on our work at various conferences, workshops and seminars and in conversations have helped us to refine our ideas and understand the viewpoints of others. Their input throughout the development of this book has been of immense value. Thanks are also due to colleagues at Dundee Law School for their very helpful comment, criticism and support throughout the project underlying this book. In particular, we are grateful to Professor Andrea Ross and Ms Yvonne Evans for their comments on drafts of some of the chapters.

Professor Reid would particularly like to thank Dr Nsoh for proving to be such a good partner in this project, especially in view of the very unsettling beginning to his work at Dundee noted above. Thanks are also due to my daughter Christine (who was ensnared when a visit home coincided with the need for some text to be checked; my other daughter Mary escaped) and especially to my wife Anne for her assistance on scientific matters and in so many other ways.

Dr Nsoh would particularly like to thank colleagues, both former and present, for their collegial assistance and support in the process of preparing this book. Special appreciation is reserved for Professor Reid, first, for employing me to work on this project and for his indulgence and understanding as I had to cope with the disruption arising from my move to start a career as a Lecturer in Law, the subsequent move and other commitments in the intervening period. Above all, thanks for his always insightful and immensely knowledgeable comments; I have benefited a tremendous amount working with him during these years and for that I will always be indebted beyond measure. The greatest thanks though are reserved for my wife Ruth, and my daughters Miya and Lina. The push to produce this work has meant many late nights and weekends and I appreciate how much of a burden this has imposed on you all.

1. Introduction

It is vital for the future of humankind that we maintain and enhance the health of the world's ecosystems and their ability to support the human population. Preventing and reversing the devastating loss of biodiversity arising from human activity is an essential objective and one that has been accepted by the governments of the world. Yet we are not being very successful at meeting this goal. This assessment was confirmed in *Global Biodiversity Outlook 3*,[1] prepared in 2010 under the auspices of the Convention on Biological Diversity:[2]

> The target agreed by the world's Governments in 2002, 'to achieve by 2010 a significant reduction of the current rate of biodiversity loss at the global, regional and national level as a contribution to poverty alleviation and to the benefit of all life on Earth', has not been met. There are multiple indications of continuing decline in biodiversity.

This gloomy conclusion of international experts reveals the need to do better. Since then, a new set of objectives has been agreed, again calling for reductions in the rate of habitat loss and extinctions and for states to ensure that concern for biodiversity is integrated into their strategies and planning processes.[3] Yet, although there are some signs of hope,[4] the overall picture remains unpromising. The latest *Global Biodiversity Outlook 4* in 2014 shows progress with regard to formulating conservation and policies and designating protected sites, but inadequate, if any,

---

[1] Secretariat of the Convention on Biological Diversity, *Global Biodiversity Outlook 3* (Montreal, 2010) 9.

[2] Convention on Biological Diversity (adopted 5 June 1992, entered into force 29 December 1993) 1760 UNTS 79.

[3] The Aichi Biodiversity Targets for 2011–20, within the Strategic Plan for Biodiversity 2011–20 agreed at the Conference of the Parties under the Convention on Biological Diversity in 2010: COP 10 Decision X/2.

[4] E.g. the biodiversity indicators for the United Kingdom do show areas where there has been positive progress in recent years (albeit in many cases not yet making up for losses in past decades), as well as others where a negative trend is continuing; Joint Nature Conservation Committee, *UK Biodiversity Indicators 2014* (JNCC, 2014).

progress in crucial areas such as using natural resources within ecological limits, preventing habitat loss, fragmentation and degradation, and preventing extinctions.[5]

Against this background it is inevitable that there should be some reassessment of how we go about trying to halt the loss of biodiversity. Achieving the goals that have been set will require enhancing the legal tools that can assist the conservation of habitats and species. Much could be achieved by a strengthening of the established approaches, which rely heavily on direct 'command and control' regulation to prevent harmful activities. Thus stronger laws could be introduced (and enforced) to prevent the disturbance of protected species, to create and protect nature reserves and to require landowners to manage their land with a view to ensuring that habitats for flora and fauna are maintained in, or restored to, a healthy condition. There is, though, growing interest in a wider range of mechanisms that utilise different approaches to support conservation. These can be grouped under the heading the 'privatisation of biodiversity', because although they may not involve a wholesale transfer of property or control into private hands (converting public goods into private goods), they do not rely on direct regulation by the state but rather make use of private rights and market devices to achieve their goals. Such approaches, as a supplement or alternative to direct regulation, are already a feature of the legal response to other environmental problems, such as the emissions trading schemes developed in response to climate change,[6] and their expansion to biodiversity issues creates both opportunities and risks.[7]

The mechanisms discussed under this broad heading of 'privatisation' are diverse, with greater or lesser elements of a genuine market-based approach,[8] but they share an emphasis on private law frameworks, private choice, private initiative and private funding as opposed to conservation being overwhelmingly an enterprise directed and controlled by public

---

[5] Secretariat of the Convention on Biological Diversity, *Global Biodiversity Outlook 4* (Montreal, 2014); see also Derek P. Tittensor and others, 'A mid-term analysis of progress toward international biodiversity targets' (2014) 346 *Science* 241.

[6] Scott D. Deatherage, *Carbon Trading Law and Practice* (OUP, 2012).

[7] Colin T. Reid, 'Between priceless and worthless: challenges in using market mechanisms for conserving biodiversity' (2013) 2 *Transnational Environmental Law* 217.

[8] Romain Pirard, 'Market-based instruments for biodiversity and ecosystem services: a lexicon' (2012) 19–20 *Environmental Science and Policy* 59; see section 1.6 below.

authorities. At the core of consideration lies payment for ecosystem services (PES) and the use of biodiversity offsets and conservation covenants (see Box 1.1), but other devices are also considered, such as use of the tax system. Covenants are based on private law obligations in relation to land ownership and may or may not operate in conjunction with other techniques such as offsetting and PES where the underlying concept firmly has its roots in market-based thinking. Such mechanisms are not wholly novel and are in use to some extent in various jurisdictions around the world. Learning from that experience is important, but lessons must be drawn with care since the differences in the physical, legal, regulatory, financial and cultural contexts across different countries mean that the scope for particular mechanisms to operate successfully in new settings varies greatly.

BOX 1.1 DEFINITIONS

- Payment for ecosystem services (see Chapter 3)
  Land managers receive payments to reflect the benefits that their land is providing to the wider environment and community, for example, water management or habitat for pollinating insects, thereby creating an incentive for them to maintain or enhance these services whilst those who benefit contribute to the cost.
- Biodiversity offsetting (see Chapter 4)
  Development causing a loss to biodiversity in one place is allowed to proceed so long as ecological gains are achieved elsewhere, thereby ensuring no net loss to biodiversity.
- Conservation covenants (see Chapter 5)
  (Also known as conservation easements or burdens or servitudes.) Landowners, freely or in exchange for payment, make long-term agreements which control the future use and management of the land, by them and by their successors in title, either as a conservation tool in its own right or as the legal basis for delivering an offset or payment scheme.

This chapter provides a brief introduction to the terminology of biodiversity and ecosystem services before examining the ways in which the law has traditionally dealt with flora and fauna and the more recent emergence of specific laws designed to conserve nature. Wider trends in environmental law and regulation are then discussed, with specific reference to the use of market-based instruments and the debates over commodification that they inspire.

1.1 BIODIVERSITY

In recent years, nature conservation has been dominated by the term 'biodiversity'. It owes its prominence to the Convention on Biological Diversity agreed at the 'Earth Summit' in Rio de Janeiro in 1992,[9] but some care must be taken in its use. The Convention provides the following definition:

> 'Biological diversity' means the variability among living organisms from all sources including, *inter alia*, terrestrial, marine and other aquatic ecosystems and the ecological complexes of which they are part; this includes diversity within species, between species and of ecosystems.[10]

The power of the definition is that it covers several different but interconnected forms of diversity, all of which must be considered if the natural heritage of the earth is to be passed down to future generations. Diversity between species, that is, the presence of a number of distinct sorts of plant or animal, is perhaps the simplest aspect of this and is what drives the concern to prevent the extinction of endangered plants and animals. Diversity within species reflects the extent of genetic variation within each species, and the fact that local populations may have traits different from those in other areas, traits that may prove essential to the long-term survival or value of the species (such as disease resistance or adaptation to climate variations) and in some cases may eventually lead to differentiation as a separate species. Diversity of ecosystems is the basis for the conservation of different habitat types, whilst diversity within ecosystems is often a measure of the health of a habitat and the absence of human interference. Nature conservation must take account of all of these aspects of our biological heritage.

Yet there is a danger that diversity by itself becomes the sole yardstick for measuring the health and value of a habitat or ecosystem. This would lead to those habitats with comparatively little diversity being under-valued, prioritising wetlands and rainforests over deserts and polar regions. Indeed, habitats with less diversity may require extra attention since they may be more vulnerable than those with greater diversity and hence greater resilience. In habitats of low species richness there is a greater risk that disruption to one species will cause the whole ecosystem to become dysfunctional, whereas where there is high species richness there is likely to be some duplication of ecological functions and hence a

---

[9] See Convention on Biological Diversity (n. 2).
[10] *Ibid.* Art. 9.

greater ability to maintain the ecosystem in the face of disturbance.[11] Moreover, taken to an extreme, a focus on numerical diversity alone may lead to surprising results. Given the diversity that can be found in suburban gardens as a result of the variety of plants and habitats that are introduced,[12] converting native grassland or many other habitats to suburban development may well increase the biodiversity of the area if all that is considered is the number of species present.

Biodiversity, however, is often used not as a strict technical term but as a catch-all phrase for 'nature'. Terminology here is difficult[13] since on close examination most common and convenient words have difficulties and reflect deeper inconsistency in our attitudes. Even in the most intensive man-made agricultural systems there are still 'natural' processes going on and some species (beyond those being deliberately cultivated) will benefit from the new ecosystem that has been created. We want to distinguish between what humans have created and the 'natural world', but that phrase itself ignores the fact that most parts of the world have been heavily affected by human intervention over centuries, if not millennia. There are very few truly 'natural' habitats left and many environments created by humans are now valued for their contribution to the beauty and diversity of fauna, flora and landscapes, and require active management to maintain their current features.[14] The term 'flora and fauna' again fails to distinguish between the 'naturally' occurring species and those introduced or promoted by humans. A further possibility is to speak of conserving[15] 'wildlife' but this ignores the geological and physiographical features that are also part of ecosystems (and many

---

[11] Rick D. Stuart-Smith and others, 'Integrating abundance and functional traits reveals new global hotspots of fish diversity' (2013) 501 *Nature* 539.

[12] A detailed study over 15 years has recorded 1,782 species of animal and 42 species of plant in one suburban garden in Leicester (Jennifer Owen, *The Ecology of a Garden: The First Fifteen Years* (CUP, 1991)), whilst over a similar period a less intensive study of a garden in Dundee has led to the recording of seven species of bumblebee, 52 hoverflies and 174 moths (data from Anne Reid; personal communication, 25 October 2015).

[13] This is a pervasive issue and it is instructive that the EU No Net Loss Working Group produced a glossary of almost 40 pages to define and explain key terms in their deliberations; see http://ec.europa.eu/environment/nature/biodiversity/nnl/pdf/NNL_Glossary.pdf.

[14] E.g. ensuring the continuing grazing of flower- and insect-rich meadows to prevent them being taken over by scrub.

[15] There is a separate terminological issue here over the evolution from the purely defensive 'protection' or 'preservation' of nature to the more positive task of 'conservation' and now even 'enhancement'.

people in thinking of the 'wildlife' of an area have in mind only its animals, not its plants, far less the microbial and bacterial elements which are also essential parts of any ecosystem). 'Natural heritage' is another term that is used[16] but does not have such wide recognition and can be seen as carrying specific legal connotations.[17] Despite such difficulties, most of the time it is clear enough what is being talked about and the marginal difficulties in priorities and scope arising from the use of specific terms can be left to be discussed in the specific contexts where they arise. Accordingly this book follows the general usage of what can strictly be viewed as loose terminology, except where specific points are to be made.

1.2 ECOSYSTEM SERVICES

Further terms which are important to the issues covered in this book are 'ecosystems' and 'ecosystem services'. Conservation today is increasingly adopting an ecosystem approach rather than focusing narrowly on particular species or habitats. This approach involves 'a strategy for the integrated management of land, water and living resources that promotes conservation and sustainable use in an equitable way', with a priority being the 'conservation of ecosystem structure and functioning, in order to maintain ecosystem services'.[18] The focus is thus shifting from a concentration on individual sites and species to a concern for the health of wider ecosystems, requiring attention to the more diffuse impacts on the wider landscape and environment rather than concentrating on the establishment and protection of discrete citadels for designated forms of wildlife.

The various elements of an ecosystem operate in ways that contribute to the health of the environment as a whole, but the valuable services that are being provided for humanity have generally been overlooked in the past. As well as efforts to increase our understanding of the working of ecosystems generally, attention is being paid to identifying and seeking recognition for

---

[16] E.g. Natural Heritage (Scotland) Act 1991, where it is provided that '"the natural heritage of Scotland" includes the flora and fauna of Scotland, its geological and physiographical features, its natural beauty and amenity' (s. 1(3)).

[17] Cf. the use of the concept 'common heritage of mankind' in the United Nations Convention on the Law of the Sea (adopted 10 December 1982, entered into force 16 November 1994) 1833 UNTS 3, and the discussion on the status of water in Chapter 8, section 8.3.

[18] Fifth Ordinary Meeting of the Conference of the Parties to the Convention on Biological Diversity, 15–26 May 2000, Nairobi, Kenya (CBD-COP 5), Decision V/6, para. 5, Principle 5.

the ways in which the natural environment contributes to human well-being. This is not a wholly uncontroversial exercise,[19] but such efforts reveal that land left in its 'natural' condition is not actually unproductive, as it is often perceived, but is providing a range of services of great practical, economic and spiritual value to society. Studies at an international level have identified many such contributions: regulating services (including filtration of pollutants by wetlands, climate regulation through carbon storage, pollination and protection from disasters); cultural services (including recreation, spiritual and aesthetic values); provisioning services (including wild foods, crops, fresh water and plant-derived medicines); and supporting services (including soil formation and nutrient cycling).[20]

BOX 1.2 ECOSYSTEM SERVICES

Ecosystem services are the benefits people obtain from ecosystems. These include provisioning, regulating and cultural services that directly affect people and supporting services needed to maintain the other services.

Provisioning services *Products obtained from ecosystems* • food; • fresh water; • fuelwood; • fibre; • biochemicals; • genetic resources.	**Cultural services** *Non-material benefits obtained from ecosystems* • spiritual and religious; • recreation and ecotourism; • aesthetic; • inspirational; • educational; • sense of place; • cultural heritage.
Regulating services *Benefits obtained from regulation of ecosystem processes* • climate regulation; • disease regulation; • water regulation; • water purification; • pollination.	**Supporting services** *Services necessary for the production of all other ecosystem services* • soil formation; • nutrient cycling; • primary production.

Source: Reformatted from UNEP Millennium Ecosystem Assessment, *Ecosystems and Human Well-Being: Synthesis* (Island Press, 2005) 57.

---

[19] Matthias Schröter and others, 'Ecosystem services as a contested concept: a synthesis of critique and counter-arguments' (2014) 7 *Conservation Letters* 514.

[20] The Economics of Ecosystems and Biodiversity (TEEB), *Mainstreaming the Economics of Nature: A Synthesis of the Approach, Conclusions and Recommendations of TEEB* (2010) 7, Annex 2, and the sources referred to there. See generally the website of the TEEB project, http://teebweb.org.

Identifying these services and those who provide and benefit from them, lies at the heart of many of the economic approaches discussed later in this book. This can be a challenging task, especially since the services are often overlapping and hard to disentangle; maintaining a healthy peat bog provides benefits for both biodiversity and carbon storage, whilst contributing with many other elements in the wider catchment to the health and quantity of water resources. Giving the services a value, especially in monetary terms, both emphasises their importance and provides a basis for creating mechanisms for attributing costs and benefits between parties in a way that goes beyond the prevalent regulatory approaches. The basic concept, though, rests on showing how humanity benefits from nature. There is plenty of scope for disagreement over what should be included, depending on how immediate or indirect the links with human benefits can be,[21] and a consequence may be that the same habitat is valued very differently depending on its location: a water meadow that absorbs flood water may have more or less value depending on whether it is upstream or downstream of a town. Moreover the focus on the provisions of services to humankind leaves this approach open to criticisms of anthropocentricity as discussed in Chapter 8, further downgrading the non-economic arguments for conservation.[22]

1.3 LAW AND NATURE

The law has been involved with the natural world since the earliest times[23] and many of the key elements of the treatment of nature in western legal systems can be traced back to Roman law. Every legal system has its own detailed rules, but the general approach of modern western systems has been to treat nature simply through the application of the rules of property. Land, plants and animals are regarded not as part of an integrated ecosystem but as individual items of property held by

---

[21] E.g. it can be argued that as well as considering the ecosystem services being delivered today there should also be consideration of 'evosystem services', placing a focus on maintaining the long-term health of ecosystems so that they can continue to evolve to provide services in the future; Daniel P. Faith and others, 'Evosystem services: an evolutionary perspective on the links between biodiversity and human well-being' (2009–10) 2 *Current Opinion in Environmental Sustainability* 66.

[22] Kent H. Redford and William M. Adams, 'Payment for ecosystem services and the challenge of saving nature' (2009) 23 *Conservation Biology* 785.

[23] G.R. Driver and John C. Miles, *The Babylonian Laws* (OUP, 1952) vol. I, 150–54.

and wholly at the disposal of their owner, with no regard for their wider environment, except to the extent that there might be interference with equivalent legal rights held by other people. Where things are not the subject of property rights, they fall outwith legal consideration altogether.

For land this means that the landscape has been fragmented into different plots which are treated quite separately, ignoring their role as part of an integrated ecosystem. Land is divided into units on the basis of boundaries which usually have no link to natural features. Even where such features are used, they are as likely to split ecosystems (for example, by using a river as a boundary) as to respect them (by using a watershed). Each unit can be managed individually with no regard for the consequences on the natural world beyond its boundaries, except to the extent that there is an infringement of property rights held by others. Although now almost invariably overlain by at least some statutory limitations imposed for a range of social goals,[24] the starting point is that the landowners are free to do what they wish with their land, so long as they do not cause a nuisance to or otherwise interfere with the rights of their neighbours.

Plants are also treated simply as pieces of property. As accretions to the soil,[25] they belong to the owners of the land[26] who are therefore (in the absence of other restrictions) entitled to exploit or destroy them as they wish. For animals, there is a distinction between domesticated and wild animals,[27] the former being wholly items of private property throughout their lives, the latter being owned by nobody unless and until killed or captured.[28] Game and fishing rights may give their holders some legal interest in the species covered, entitling them to prevent others harming these interests, but otherwise wild animals may fall wholly outwith the law's attention, except to the extent that the state asserts a general right in them.[29]

---

[24] Such as to limit pollution.

[25] *Digest*, 41.1.7.13, 41.1.9.pr.

[26] There may be special rules to deal with the ownership of crops between landlord and tenant or when properties are sold, as illustrated in *Boskabelle Ltd* v. *Laird* [2006] CSOH 173, 2006 SLT 1079.

[27] The distinction between domesticated and wild animals includes rules of long-standing that recognise the ownership of animals such as bees which are free to roam but consistently return to their owner's premises; *Digest* 41.1.5, *Kearry* v. *Pattinson* [1939] 1 KB 471.

[28] There may be some exceptions, e.g. the rules in England conferring on the Crown rights in swans, sturgeon and whales; *Case of Swans* (1592) 7 Co. Rep. 15b, Blackstone, *Commentaries* (1783) i 223.

[29] See the discussion in Australia in *Yanner* v. *Eaton* [1999] HCA 53, [1999] 201 CLR 351; for the position in the United States see O. Houck, 'Why do we

This legal approach has several consequences. Any measures introduced to conserve or enhance habitat will amount to an interference with the rights of the owner of the land concerned and thus need specific legal authorisation. Moreover, since the right to property is often protected as a constitutional or human right, such interference must satisfy the standards set for justifying such action. Thus in the United Kingdom, the Human Rights Act 1998 gives effect to the European Convention on Human Rights which in Article 1 of Protocol 1 includes the right to the 'peaceful enjoyment of ... possessions'. This has been used by landowners to challenge the legitimacy, in the absence of compensation, of the imposition of habitat conservation measures which they claimed adversely affected their property rights. Environmental protection, including nature conservation, has been accepted by the European Court of Human Rights as one of the grounds on which the owners' rights can legally be restricted[30] and the English courts held that the measures in question fell far short of the point where there was a disproportionate and thus unlawful interference with the owners' rights.[31] In other jurisdictions that boundary may be set at a different point, and in the United States, the concept of 'regulatory takings' is likely to permit the state much less freedom of action. The law in the United States operates on the basis that the government can deprive owners of their property not just by direct expropriation but also by intrusive regulation which removes the owners' freedom to use the property as they wish and thus its effective value.[32] Expropriation by the state is lawful only if compensation is provided,[33] and in the argument over the point at which mere regulation (for conservation or other purposes) becomes a 'taking' (and therefore

---

protect endangered species, and what does that say about whether restrictions on private property to protect them constitute takings' (1994–95) 80 *Iowa Law Review* 297 and the recently reinvigorated debates about the public trust doctrine (see Chapter 2, section 2.9).

[30] *Fredin* v. *Sweden* (1991) 13 EHRR 784.

[31] *R (Trailer and Marina (Leven) Ltd)* v. *Secretary of State for the Environment, Food and Rural Affairs* [2004] EWCA Civ 1580, [2005] 1 WLR 1267.

[32] There is a substantial, if not always straightforward, volume of case law on when regulation of property amounts to its 'taking', e.g. *Penn Central Transportation Co.* v. *New York City*, 438 US 104 (1978); *Lucas* v. *South Carolina Coastal Council*, 505 US 1003 (1992); *Koontz* v. *St Johns River Water Management District,* (Docket no. 11-1447) 570 US _ (2013), 133 S Ct 2586 (2013).

[33] US Constitution, Fifth Amendment.

unlawful unless compensation is provided),[34] the threshold is set more in the owners' favour than is the case in some other countries, such as the United Kingdom.

The recognition of ownership of plants and animals gives the owners a clear entitlement to take steps to protect their property against external threats. This enables them to take measures against the actions of others, and also against natural processes, which risk damaging or destroying that property, but does not require them to respond to take any protective action – indeed allowing, or even causing, the destruction of one's own property is one of the most significant rights that an owner has. At the same time, though, this ownership excludes other parties from having any legal interest in the condition, or even survival, of the property owned by another person. For wild animals that are not owned, the fact that nobody has a legally recognised interest in them further restricts the extent to which any concern for them can be reflected in the law. Game or fishing rights, held along with or separate from ownership of land, do create a limited legal interest in the condition of certain species, but clearly this will be of limited application.[35] Generally, therefore, unless they fall within the scope of property rights which the owner chooses to protect, the welfare of the environment and of its wildlife are not matters of legal concern. It is against this legal background that legislative intervention to support biodiversity has had to operate.

1.4 NATURE CONSERVATION LAW

Legal measures to protect elements of the natural environment have existed for centuries and the nineteenth century saw the start of a shift in focus from simply ensuring the sustainability of natural resources to be exploited in future (as in the mediaeval game and hunting laws),[36]

---

[34] For an argument that there should be compensation for landowners affected by the Endangered Species Act, see N. Scott Arnold, 'The Endangered Species Act, regulatory takings, and public goods' (2009) 26 *Social Philosophy and Policy* 353.

[35] See Chapter 7, section 7.3.

[36] John M. Gilbert, *Hunting and Hunting Reserves in Medieval Scotland* (John Donald, 1979); Colin T. Reid, 'Environmental legislation of the Scottish Parliament' in Hector L. McQueen (ed.), *Stair Society: Miscellany Six* (Stair Society, 2009) 65–9.

towards providing for protection against cruelty and the wanton destruction of wildlife (as in the Sea Birds Preservation Act 1869 and its successors in the United Kingdom),[37] and for the preservation of exceptional areas of natural beauty (as in the recognition of what were to become the National Parks in the United States).[38] The modern law of nature conservation has taken shape during the last few decades of the twentieth century, with a wealth of international and national legislative initiatives.

The law has gradually become stricter and more far-reaching, offering more effective protection to a wider range of biodiversity,[39] but remains dominated by measures aimed to prevent harm to particular species and habitats. In both cases the basis is that the state identifies features that are considered valuable and then imposes restrictions on activities that might harm them. For species, the approach has been to identify a number of animals (including birds, fish and invertebrates) which are to be protected and then make it a criminal offence to kill or capture them or to harm them in other ways. For habitats, again the starting point has been to identify and designate the areas of land deserving protection and then to seek their conservation. Initially, this again involved creating criminal offences prohibiting any direct harm to the sites, but more recently these provisions have been accompanied by measures to support the positive conservation and enhancement of the designated areas. The details of the law will vary between jurisdictions in the number and variety of species and habitats protected, the scope of the protection offered[40] and the ancillary measures adopted to support and enforce the main provisions (for example, offences of selling or simply possessing specimens of

---

[37] There were 12 statutes passed in the United Kingdom in relation to wild birds between 1869 and 1908.

[38] An Act to set apart a certain Tract of Land lying near Head-waters of the Yellowstone River as a public Park; 1872 ch. 24 (*US Statutes at Large*, vol. 17, 32).

[39] For a discussion of the evolution of the law in one jurisdiction see Colin T. Reid, 'Towards a biodiversity law: the changing nature of wildlife law in Scotland' (2012) 15 *Journal of International Wildlife Law and Policy* 202.

[40] The categories of protected areas recognised by the IUCN is one illustration of the different levels of and objectives for habitat protection; see Alexander Gillespie, *Protected Areas and International Environmental Law* (Martinus Nijhoff, 2007) ch. 2.

protected species), but these key features lie at the core of international,[41] regional[42] and national[43] conservation law.

Thus in Scotland,[44] it is an offence intentionally or recklessly to kill, injure or take any wild bird, or to take eggs or destroy nests, whilst for certain species the offences include causing disturbance while nesting. Well over 100 species of wild animals (including fish and invertebrates) are similarly protected and it is an offence intentionally or recklessly to pick, uproot or destroy any of over 200 species of plants (including mosses and liverworts), or to pick or destroy their seeds or spores. The possession or sale of specimens of the protected species is also an offence. There are several designations of habitat, attracting different levels of protection, but for those designated as European Sites or Sites of Special Scientific Interest[45] it is an offence for anyone intentionally or recklessly to damage any of the natural features which have led to the site's designation, whilst an individually determined list of activities likely to damage the site is prohibited without express permission from the statutory nature conservation body. For European Sites, any plan or project which may have an adverse impact on the integrity of the site must be prevented unless certain very strict criteria are met and compensatory action taken. Throughout, there are detailed exceptions and provision for licences to be granted to permit otherwise prohibited

---

[41] E.g. the protection of migratory species under the Convention on the Conservation of Migratory Species of Wild Animals (adopted 23 June 1979, entered into force 1 November 1983) 1651 UNTS 333, and of world heritage sites under the UNESCO Convention Concerning the Protection of World Natural and Cultural Heritage (adopted 16 November 1972, entered into force 17 December 1975) 1037 UNTS 151.

[42] E.g. within the European Union in the Birds and the Habitats Directives: European Parliament and Council Directive 2009/147/EC of 30 November 2009 on the conservation of wild birds (codifying the amendments to Directive 79/409/EEC), [2010] OJ L20/7 and Council Directive 92/43/EEC of 21 May 1992 on the conservation of natural habitats and of wild fauna and flora, [1992] OJ L206/7.

[43] For the law in the United Kingdom see Colin T. Reid, *Nature Conservation Law* (3rd edn, W. Green, 2009) and Christopher P. Rodgers, *The Law of Nature Conservation: Property, Environment and the Limits of Law* (OUP, 2013).

[44] This is, of course, only a partial and greatly oversimplified account of a mass of detailed legislation. For more detail see Reid, 'Towards a biodiversity law' (n. 39) and *Nature Conservation Law* (n. 43) and Colin T. Reid, 'Nature conservation' in Frank McManus (ed.), *Environmental Law in Scotland* (W. Green, 2010).

[45] These are the main designations of protected habitat under EU and British legislation.

activities. A number of measures to support the maintenance and enhancement of habitat also exist, such as management agreements which can include payments to landowners for work that they carry out to benefit biodiversity; in practice these are the preferred method of intervention, but against a background of stricter legal controls. A statutory body, Scottish Natural Heritage, is entrusted with major powers in relation to the designation of sites and species and the enforcement of the law.

The core of the law is thus based on the traditional approach of identifying harms to be avoided and following the classic model of 'command and control' regulation: 'That is, legislatures have proscribed certain behaviour and set up a regulatory agency to monitor and police compliance with legal standards'.[46] With stricter controls being imposed as time has progressed, both helping to shape as well as reflecting changes in social attitudes, this approach has achieved some notable successes. In Great Britain, the collecting of eggs from birds' nests has been transformed from an acceptable hobby into behaviour that attracts a prison sentence and public opprobrium. Yet biodiversity continues to suffer. There is no doubt that prohibitions supported by the criminal law will continue to play a major part in conservation law, but there is growing interest in a wider role for new mechanisms.

Referring to the mechanisms and approaches discussed in this book as 'new' is a useful short-hand but not wholly accurate. As is made clear throughout this book, around the world there are many schemes currently in operation that apply the methods explored below. Many areas dedicated to conservation are in private ownership and conservation is supported through a range of schemes.[47] The management agreements which play an important part in looking after habitat in the United Kingdom are a form of payment for ecosystem services (based on something akin to a short-term conservation covenant); biodiversity offsetting and bio-banking schemes are established in Australia and the United States and conservation easements are used across the United States.[48] Nevertheless, even in such contexts, such mechanisms are a more recent addition to the more traditional 'command and control' approach. It is the contrasts between that traditional approach and the

---

[46] Neil Gunningham and Peter Grabosky, *Smart Regulation: Designing Environmental Policy* (OUP, 1998) 5, see also 38–50.

[47] Jeffrey A. Langholz and Wolf Krug, 'New forms of biodiversity governance: non-state actors and the Private Protected Area Action Plan' (2004) 7 *Journal of International Wildlife Law and Policy* 9.

[48] See Chapters 3, 4 and 5, respectively.

'new' one, and the scope for the latter to move from the margins to take a more central role in conservation law, that are at the heart of the analysis here.

The desire to explore such newer mechanisms and examine whether they can play a larger role is part of a wider movement in environmental regulation towards a more mixed 'toolkit'. To some extent this has been driven by ideological considerations as discussed below, but there is also a recognition of the limitations of the established approach. In the western world, laws such as those described above have been reasonably effective in achieving protection against direct harm for the elements of biodiversity singled out for attention, but this leaves major gaps. Species are protected once they are recognised as rare or endangered, but little may be done during the early slide from abundance to rarity, whilst areas of protected habitat risk becoming isolated citadels of biodiversity sitting in an increasingly degraded and hostile countryside. Flora and fauna suffer huge losses not as the result of any specific actions directed against them but because of the steady loss of land to intensive agriculture and development and through the cumulative effects of diffuse pollution of many sorts. The potential of new approaches to extend the impact of conservation law to address the continual erosion of biodiversity in such ways needs to be considered alongside the re-evaluation of current approaches from the perspective of regulatory theory.

1.5 ENVIRONMENTAL REGULATION

Across environmental issues a regulatory approach based on 'command and control' as described above was the dominant one as societies began to take environmental concerns seriously in the latter half of the twentieth century,[49] but it has come under significant challenge. As Hawkins wrote in 1998:

> Twenty or thirty years ago we were secure in our assumption that the most appropriate legislative response to the undesirable consequences of industrialised life for the natural environment was the use of legal regulation backed by the criminal sanction. This reliance upon command and control regulation

---

[49] Environmental laws are, of course, much older but initially tended to have a narrower focus on local and public health issues; Stuart Bell, Donald McGillivray and Ole W. Pedersen, *Environmental Law* (8th edn, OUP, 2013) 18–24.

was, however, soon to be threatened by a growing belief that it was both ineffective and inefficient.[50]

In its place has come a range of new approaches, many building on economic analyses of the position.[51] As discussed below, there are significant differences between biodiversity and the greenhouse gases which are the subject of the most strongly market-based mechanisms such as emissions trading or offsetting, but it is inevitable that attention should turn to applying the innovative techniques used for other environmental problems to biodiversity, especially since the existing approach is not achieving its goals.

The evolution of environmental regulation is a vast topic in itself,[52] but two strands have lain at the heart of recent developments. The first has been criticism of the rigid, bureaucratic and inefficient nature of 'command and control' regulation, which is seen as imposing excessive regulatory burdens on economic activity and personal freedom. Some of this criticism has been misplaced since it has targeted a stereotyped view of traditional regulation without reflecting the variety and innovation which can make this approach much more responsive. Nevertheless, there remain concerns based on the aspects just mentioned as well as the lack of incentives for going beyond the strict regulatory requirements, the adversarial nature of much enforcement action and the fact that the system operates on the basis that 'the regulator knows best', even in complex areas where information is scattered and interpretations and priorities vary.[53]

The second element has been the wider ideological tide towards deregulation and market-based approaches. The neo-classical economic analysis of environmental problems as a result of market failures has led to responses based not on direct regulatory intervention but on solutions that are themselves market-based. Strengthening private property rights so that resources are cherished, as propounded in Hardin's famous work on 'The Tragedy of the Commons',[54] is seen as the way of preventing the

---

[50] Keith Hawkins, 'General Editor's Introduction' to Gunningham and Grabosky, *Smart Regulation* (n. 46) vii.

[51] Neil Gunningham, 'Environment law, regulation and governance: shifting architectures' (2009) 21 *Journal of Environmental Law* 179.

[52] See e.g., the text and works extracted in Elizabeth Fisher, Bettina Lange and Eloise Scotford, *Environmental Law: Text, Cases, and Materials* (OUP, 2013) ch. 12.

[53] For a summary of strengths and weaknesses of different regulatory mechanisms see Gunningham and Grabosky, *Smart Regulation* (n. 46) ch. 2.

[54] Garrett Hardin, 'The tragedy of the commons' (1968) 162 *Science* 1243.

degradation of shared resources, whilst reducing the role of the state[55] has been at the heart of conservative political thinking, especially in the United Kingdom and United States. This has led to a preference for relying on mechanisms other than direct regulation to achieve policy goals. Taxonomies and labels vary and categories overlap,[56] but in addition to direct regulation the instruments used have included self-regulation, educational and information strategies and especially economic instruments and free market initiatives.[57] The initial diversification has been in pollution control and examples abound of voluntary schemes, educational programmes aimed at changing environmentally harmful practices, environmental taxes, emissions trading and offsetting schemes. Not all such innovations have been successful and the failures offer lessons to be taken into account in designing further schemes.[58] Inevitably, though, the question of expanding the reach of such approaches to include biodiversity has been raised, especially in relation to market-based instruments.[59]

1.6 MARKET-BASED INSTRUMENTS

In the same way as the term 'biodiversity' is ubiquitous in discussion about conserving ecosystems but can be used to cover different concepts, the term 'market-based instruments' also appears throughout discussion of regulatory styles, but with a range of meanings. As Pirard notes:

> [T]he instruments presented in the literature as 'market-based instruments' (MBIs) constitute an extremely heterogeneous group with loose and contrasted links to markets as defined by economic theory … In fact, the market

---

[55] Underestimating, at times, the major role that the state must play in shaping the market in areas where commerce has not naturally developed.

[56] '[I]t is better to think of the following instrument categories as points on a compass rather than discrete or absolute policy units.' Gunningham and Grabosky, *Smart Regulation* (n. 46) 38

[57] OECD, *Environmental Policy: How to Apply Economic Instruments* (OECD, 1991).

[58] See the examples given in Robert L. Glicksman and Thoko Kaime, 'A comparative analysis of accountability mechanisms for ecosystem services markets in the United States and the European Union' (2012) 2 *Transnational Environmental Law* 259, 266–70.

[59] OECD, *Harnessing Markets for Biodiversity: Towards Conservation and Sustainable Use* (OECD, 2003).

terminology seems to have been adopted by default, as a way of differentiating these instruments from all other approaches that do not have a pricing element.[60]

The uncertainty over terminology is exacerbated by the fact that there can be many variations of the different mechanisms, involving different levels of 'true' market elements. For example, at its extremes a biodiversity offsetting scheme can involve no market or trading element, requiring developers themselves directly to provide habitat equivalent to that being lost, or can merge into a bio-banking scheme based on credits which are freely traded on a market basis.[61] Agonising over the precise taxonomy can be a distraction from considering the practical benefits and dangers of the various mechanisms and how they can be used. Nevertheless, the different categories of instrument do embody different relationships between parties and serve different goals, so that appreciating the variety can be significant in structural discussions.

Before considering this, though, it is useful to identify the advantages claimed in the conservation context for the approaches included under the extensive heading 'market-based instruments'.[62] Three main arguments are the correction of market failures, the provision of incentives and the provision of funding beyond that otherwise available for conservation purposes.[63] At its simplest, the idea of market failure centres on the recognition that the existing relationships between parties do not recognise adequately, if at all, the costs and benefits arising from various interactions. The market prices for goods or services do not recognise the externalities involved in various processes. For example when manufacturers can freely discharge polluted water into a river there is no financial cost to them and so no impact on the prices they charge to consumers,

[60] Romain Pirard, 'Market based instruments' (n. 8) 19–20 *Environmental Science and Policy* 59, 66; see also Erik Gómez-Baggethun and others, 'The history of ecosystem services in economic theory and practice: from early notions to markets and payment schemes' (2010) 69 *Ecological Economics* 1209.

[61] Biodiversity offsetting is discussed in Chapter 4.

[62] 'Applying economic thinking to the use of biodiversity and ecosystem services can help clarify ... why successful environmental protection needs to be grounded in sound economics, including explicit recognition, efficient allocation, and fair distribution of the costs and benefits of conservation and sustainable use of natural resources': TEEB, *Mainstreaming the Economics of Nature* (n. 20) 3; see generally http://teebweb.org.

[63] Pirard, 'Market based instruments' (n. 8); Romain Pirard and Renaud Lapeyre, 'Classifying market-based instruments for ecosystem services: a guide to the literature jungle' (2014) 9 *Ecosystem Services* 106.

yet very real disadvantages are borne by those who use in various ways the rivers which now suffer poorer water quality. Market instruments can be used to ensure that the burdens currently borne by others are in fact attributed to and carried by those whose activities create them, that 'the polluter pays'. Similarly, the externality can be a positive one, where individuals remain unrewarded for services they are providing and from which other members of society benefit, such as when a derelict eyesore in a suburban area is renovated, generating aesthetic benefits and increased house prices for neighbours.

The spread of such thinking into the area of biodiversity has been aided by the development of the notion of ecosystem services. The many benefits that society gains from 'undeveloped' land have been recognised, but also the absence in most cases of any means by which the suppliers of these services are rewarded by the beneficiaries. Providing these services entails costs, whether management expenses or income foregone by not converting land to uses which are more directly rewarded, and it is unfair that the beneficiaries do not contribute to these. Market correction, or rather market creation, can provide mechanisms for such a contribution to be made, supplying an incentive to maintain or enhance the provision of the service, as opposed to the present position where the incentives are to maximise the conventionally profitable uses of land, at the expense of ecosystem services.

The second argument is also about incentives, but concentrates on a general criticism of 'command and control' regulations that, while they can succeed in stopping people from doing the specific bad things which a policy wishes to eliminate, they do nothing to encourage them to do more of the good things which should be promoted. Thus, for factory operators a rational response to pollution regulations which set a maximum allowable level of certain emissions is simply to ensure that their emissions are just below that level. There is no incentive for them to make further reductions, even though emissions below the standard set are still damaging to the environment. Similarly in a biodiversity context, although designated sites and species will benefit from general compliance with the prohibitions introduced to prevent the most obvious direct harm to them, eliminating such directly harmful activity falls well short of guaranteeing the sympathetic environment in which all elements of biodiversity can thrive.

Economic instruments can provide incentives to do more of the 'good things', whilst leaving individuals free to determine for themselves exactly how far they should go. They are thus particularly well suited to dealing with more diffuse problems, where environmental harm is the result of the cumulative impact of many individual actions. Whereas specific permits setting an annual maximum for greenhouse gas emissions might work for major site-specific installations, such an approach is less likely to be practicable in tackling the emissions caused by the millions of motor vehicles in a country, used in different ways and to different extents by millions of individuals. On the other hand, an environmental tax on petrol provides an incentive for everyone to drive less or to use a more fuel-efficient vehicle, achieving the goal of an overall reduction in emissions. In the same way, a requirement to offset any harm to biodiversity caused by development provides an incentive to avoid or limit the harm being done, providing benefits for biodiversity beyond what can be achieved by prohibitions on activity that harms the most valuable sites designated for protection.

The third argument for using market-based instruments is the potential they provide for attracting a new source of income to pay for conservation work, whether management and enhancement of existing habitat, the restoration or (re-)creation of habitat that has been degraded or the monitoring and enforcement to support measures aimed at preventing further harm.[64] The conventional pattern of regulation is based on public authorities taking the lead in identifying and supporting the features to be protected, paying (directly or through assistance to landowners) for management and enhancement and bearing all the administrative and enforcement costs. Market-based instruments can spread this cost, so that the polluter or beneficiary, not the state, is paying. At a time when the need for greater conservation efforts coincides with austerity affecting the public sector in so many countries, this may be a particularly attractive aspect when greater use of market-based instruments is considered (see Table 1.1).

The range of instruments is extensive and can be categorised in various ways. Pirard presents a 'theory-based lexicon',[65] but since, as noted above, many of the schemes in use or under consideration can display characteristics of different classes of instrument, the discussion that follows is organised on the basis of how mechanisms present themselves in the policy-maker's toolkit rather than under the sophisticated analysis of an economic theoretician. Nevertheless, being aware of the conceptual

---

64 Emma Comerford, Dominic Molloy and Paul Morling, *Financing Nature in an Age of Austerity* (RSPB, 2010); eftec, *Innovative Use of Financial Instruments and Approaches to Enhance Private Sector Finance of Biodiversity, Final Summary Report to European Commission Directorate-General Environment*, 070307/2010/581922/ETU/F1 (2012).

65 Pirard, 'Market based instruments' (n. 8).

Table 1.1 Market-based instruments

Category	*Exclusive characteristics*
Direct markets	A market where an environmental product can be directly traded between producers and consumers (or processors).
Tradable permits	An ad-hoc market where users of an environmental resource need to purchase 'permits' that can be further exchanged among resource users, thereby creating artificial scarcity.
Reverse auctions	A mechanism whereby candidates to service provision set the level of payment (if accepted) in response to a call by public authorities to remunerate landholders.
Coasean-type agreements	Ideally spontaneous transactions (free of public intervention) for an exchange of rights in response to a common interest of the beneficiary and the provider.
Regulatory price changes	Consists in regulatory measures that lead to higher or lower relative prices.
Voluntary price signals	Consists in schemes whereby producers send a signal to consumers that environmental impacts are positive (in relative terms) and consequently gain a premium on the market price.

Source: Romain Pirard and Renaud Lapeyre, 'Classifying market-based instruments for ecosystem services: a guide to the literature jungle' (2014) 9 *Ecosystem Services* 106, 109.

and structural underpinnings is valuable. In particular, some reflection on what is needed to make an effective market helps in assessing the appropriateness of a market-based approach for various ways of treating biodiversity. In so doing, the continuing role of the state in providing the conditions within which a market can operate must be appreciated, from ensuring the basic background of respect for the rule of law to the establishment of new forms of rights. This emphasises that a shift from 'command and control' to market-based instruments entails a change in the nature of the state's role in regulating an activity rather than its withdrawal from it.[66]

---

[66] OECD, *Handbook of Market Creation for Biodiversity: Issues in Implementation* (OECD, 2004).

The elements of a market-based system can be set out differently, but when one looks at the five requirements for 'a workable platform for markets' identified by McMillan,[67] all of them present challenges when dealing with biodiversity as opposed to established commercial goods. The requirement that 'information flows smoothly' requires, first, that the information exists, whereas in many cases we do not really understand the state and workings of our environment. For most species we lack data on population size and distribution and have limited knowledge of the nature and significance of all the interactions between the elements within an ecosystem. There is then the further issue of who holds the information.[68] The occupiers of the land will have detailed local knowledge of conditions and probably some awareness of the macro-flora and fauna. Conservation bodies may have specialist scientific knowledge at a general level but less site-specific knowledge. Meanwhile, none of this information is readily available to would-be market entrants. This lack of knowledge makes it difficult to ensure a second requirement, that 'side effects on third parties are curtailed', quite apart from the risk of unforeseen consequences as wholly new incentives are created to influence land managers and others. Such risks are illustrated by the belated realisation that moves to increase the use of biofuels were actually having negative social and environmental impacts by encouraging intensive cultivation of source plants in place of food production or more biodiversity-friendly land uses.[69]

McMillan's third requirement that 'property rights are protected' requires the identification of distinct rights, which in this context requires the creation of novel rights which can be legally recognised since, as noted above, biodiversity is not well served by traditional legal structures. Moreover these rights have to be defined[70] and allocated in a way that allows for them to be traded and for mechanisms to be developed for their protection. This task is linked to the fourth requirement, that 'people can be trusted to live up to their promises', something which depends not

---

67 John McMillan, *Reinventing the Bazaar: A Natural History of Markets* (Norton, 2002) ix.

68 See Chapter 2, section 2.11.

69 A. Webb and D. Coates, *Biofuels and Biodiversity*, Technical Series No. 65 (Secretariat of the Convention on Biological Diversity, 2012).

70 Callon and Muniesa analyse markets as 'calculative collective devices' and identify the need to detach distinct entities to be manipulated, far from straightforward in the inter-related world of biodiversity; Michael Callon and Fabian Muniesa, 'Peripheral vision: economic markets as calculative collective devices' (2005) 29 *Organization Studies* 1229.

just on bare trust but on the existence of mechanisms to guarantee and enforce the rights created and traded. The challenges here include the inherent artificiality of rights such as biodiversity credits, the long-term nature of any obligations to create or maintain habitat and the dynamic nature of the environment, especially in an age of climate change, which may render futile even the most carefully thought out and implemented scheme to achieve particular biodiversity goals. McMillan's final requirement, that 'competition is fostered,' calls attention to the need to limit transaction costs if any market is to succeed. It is too often forgotten that Coase's seminal article 'The Problem of Social Cost' stresses the vital importance of transaction costs, noting both that his initial analysis of the merits of using markets to rearrange rights to increase the value of production rests on the 'very unrealistic assumption' that there 'were no costs involved in carrying out market transactions' and that such costs may mean that the optimal arrangement of rights is never achieved.[71] In tackling transaction costs, thought has to be given to ways of stimulating a large enough market to achieve efficiencies of scale and those arising from repeat dealings, whilst the potential role of brokers and other intermediaries and the means of improving information flows are further aspects to be considered.

There are, therefore, challenges to be faced in designing an effective means of implementing a market-based approach, as well as the need to match the mechanism to the objective. To take the simple example of an environmental tax, if the goal is to discourage the undesirable activity, then a high tax rate should be set in order to provide a strong disincentive. On the other hand, if the goal is to raise revenue to be used for environmental (or other) purposes, the tax rate should be set at a lower level so that the taxable activity continues but the optimal revenue is raised. Given the multi-faceted nature of much environmental policy, and its interactions with social and economic concerns, there is rarely such a simple goal to be followed, quite apart from the uncertainties of exactly how any market-based instrument will operate in practice.

1.7 COMMODIFICATION

The technical challenges involved in designing effective market-based schemes and ensuring their implementation are not the only ones faced by such an approach. There is a much more fundamental one that

[71] Ronald Coase, 'The problem of social costs' (1960) 3 *Journal of Law and Economics* 1, 15–16.

questions the legitimacy of this approach at a basic ethical level. The selection of regulatory mechanisms is not value-free[72] and opting for a market-based approach involves accepting certain views about the human relationship with nature. The underlying notion that elements of biodiversity can be traded as commodities, given a monetary value, or even be subject to the standard rules of private property, are abhorrent from some perspectives. From such viewpoints, humans are not distinct from the rest of nature but part of the one 'Earth Community'[73] and market-based mechanisms such as biodiversity offsetting are 'anathema to the notion that nature is diverse, complex, varied and multiform and that has [*sic*] the power to participate in its own creation and recreation'.[74] These issues are discussed more fully in Chapter 8. Criticisms of market-based approaches derive from a variety of perspectives, some stemming from an instrumental concern over the distorting effects and limits of markets, whereas others rest on concepts of what should be seen as the common heritage of humankind, and some derive from a more spiritual concern for the Earth Community as a whole.

What these perspectives share is the view that biodiversity should not be regarded as a commodity.[75] From this viewpoint, elements of biodiversity are not interchangeable and their value cannot be captured wholly in monetary terms. Biodiversity is not something to be destroyed, replaced or created by human hands. Therefore, legal approaches which rest on valuation, exchangeability and making nature the subject of transactions are misconceived. As Sagoff has noted in a US context, 'To notice that the Endangered Species Act is not cost-beneficial is to recognize the obvious. That is the point of the Act, and of much of our environmental legislation.'[76]

At another level, there is a further mismatch between treating biodiversity in a market-based way and concentrating on what is appropriate

---

72 Jerneja Penca, 'Marketing the market: the ideology of market mechanisms for biodiversity conservation' (2013) 2 *Transnational Environmental Law* 235.

73 Cormac Cullinan, *Wild Law: A Manifesto for Earth Justice* (2nd edn, Green Books, 2011) 147–8.

74 Brendan Grigg, 'Biodiversity offsets: a dangerous trade in wildlife?' in Michelle Maloney and Peter Burdon (eds), *Wild Law: In Practice* (GlassHouse, 2014) 226.

75 For a fuller consideration of the implications of viewing something as a 'commodity' and difficulties in applying this to biodiversity see Nicolás Kosoy and Esteve Corbera, 'Payments for ecosystem services as commodity fetishism' (2010) 69 *Ecological Economics* 1228.

76 Mark Sagoff, 'Economics and environmental law' (1980–81) 79 *Michigan Law Review* 1393, 1411.

for its well-being. The traditional property-based legal treatment of land and nature does not match the ecosystem approach that is now accepted as necessary for effective conservation. The Convention on Biological Diversity defines an 'ecosystem' as 'a dynamic complex of plant, animal and micro-organism communities and their non-living environment interacting as a functional unit'[77] and the ecosystem approach calls for 'adaptive management to deal with the complex and dynamic nature of ecosystems',[78] noting that this requires attention to ecosystems at all scales, 'a grain of soil, a pond, a forest, a biome or the entire biosphere'.[79] The natural world, however, has been fragmented by the law into different elements which are treated quite separately, ignoring their role as part of an integrated ecosystem. Land is divided into units on the basis of boundaries which are usually not based on natural features. Each unit is managed in isolation, without regard to the implications for the wider ecosystem, or even other properties, unless other property rights are going to be affected.[80] The extension of property-based concepts to other elements of biodiversity, it might be argued, will only exacerbate the mismatch between the atomised approach based on private rights and the holistic view needed to reverse the loss of biodiversity.

1.8 OVERVIEW

The fact that the current legal approaches are not delivering the benefits for biodiversity that are required to secure the survival of a healthy and resilient environment means that attention is being paid to new ways of addressing the issue. The alternative mechanisms being considered present challenges at various levels. For instance, any scheme based on balancing or trading biodiversity units is making major assumptions both about the equivalence of 'units', which is not beyond question on ethical grounds, and about our ability to measure their value, which is not beyond question on scientific grounds. Such questions are explored more

---

77 Convention on Biological Diversity, Art. 2.

78 Convention on Biological Diversity COP 5 Decision V/6 (2000) (n. 18) para. 3.

79 *Ibid.*

80 The need to overcome this underpins the emerging concept of connectivity conservation; B. Lausche and others, *The Legal Aspects of Connectivity Conservation: A Concept Paper* (IUCN, 2013). This is also an area where greater legal recognition of ecosystem services might benefit conservation, recognising activities diminishing ecosystem services as a form of damage to the property that would benefit.

fully in the following chapters. Yet this is not to say that an offsetting scheme or any of the other new approaches should not proceed; the fact that they fall short of perfection is not a reason for refusing to contemplate their adoption. There may be very real gains to be made over the current position, but progress should be made with an explicit recognition of the ethical choices involved and of the limitations in assessing value and equivalence and with regard to the risks as well as the benefits of employing new techniques. Such recognition of both the potential and the peril of innovations has been a feature of recent debate about proposals in England to introduce biodiversity offsetting and conservation covenants, with stakeholders on all sides expressing the view that such a scheme might improve the present situation, but only so long as the details are right (although not necessarily agreeing on what the right details are).[81]

This conditional acceptance of the legal mechanisms discussed in this book places a particular onus both on the detailed working through of particular schemes and on finding the place for each technique within the variety of regulatory mechanisms which will have to be deployed together in any credible and effective set of biodiversity laws. The following chapters examine the specifics of different mechanisms, and the variations within each, looking at payment for ecosystem services, biodiversity offsetting, conservation covenants, taxation, transferable development permits, quotas and impact fees. Within each instance and between them, the aim is to identify not a single favoured solution, but the questions which have to be answered in designing a scheme that will meet the needs of the specific policy goals and the legal and physical context in which the mechanism is to be deployed. Before that, though, consideration is given to a number of pervasive issues which arise, in different forms and with varying intensity, across all of these mechanisms. The book ends with a more thorough consideration of the ethical questions raised by a 'privatised' approach to biodiversity, before trying to draw some brief conclusions.

---

[81] The authors have observed this first-hand at a range of workshops and conferences related to both the Government's proposals on biodiversity offsetting and the Law Commission's on conservation covenants (discussed in Chapters 4 and 5).

2. Pervasive issues

There are many different ways in which conservation law could be adjusted to allow more scope for private initiative or market-based instruments. The different mechanisms discussed in detail in the following chapters are in many ways quite distinct, although elements can be combined in various ways. Nevertheless, a number of pervasive issues arise when any of these are considered. Wherever there is an element of market thinking, there has to be clarity on what it is that is to be exchanged, sold or paid for, and whether there is to be a common currency or some other basis for exchange. How any new scheme is to be governed is important, whilst attention needs to be paid to the ways in which biodiversity is different from other aspects of the environment (such as greenhouse gases) where market-based instruments have made more progress. Since such issues underlie much of the more detailed thinking about specific schemes, it makes sense to look at them first in more general terms, reflecting on the challenges they create.[1]

These pervasive issues are themselves inter-linked, there is no ideal order of addressing them and inevitably at times they merge into one another. Moreover, the way in which one is addressed may avoid, introduce, reduce or exacerbate the complexity and significance of others. The extent to which a detailed resolution is needed will vary and people may agree or disagree on the merits of a particular detail in a particular context despite holding differing or similar views on some of the underlying ideals. There should always be an awareness that these different perspectives exist and an appreciation that taking certain decisions in designing or adopting a scheme also involves taking a position on certain underlying issues, expressly or implicitly. The design challenges are not just technical but do have deeper roots.

Highlighting the issues that should be considered can seem to be creating obstacles rather than just providing a checklist for progress. Two important points must therefore be appreciated at this stage, since reading

[1] Nick Hanley and others, 'How should we incentivize private landowners to "produce" more biodiversity?' (2012) 28 *Oxford Review of Economic Policy* 93, 105–9.

through the pages that follow can create a rather negative impression. The first is that many of these issues are not unique to the sorts of mechanisms that are the focus of this book, but arise in the design and operation of more traditional conservation approaches as well. There may be no monetary or other values attached, but selecting those sites and species which are to be designated as attracting legal protection is as much an exercise of comparative valuation as the more commercial valuations involved in market-based instruments. Secondly, although the issues need to be recognised, there is no need for perfect solutions to be found for all of them before the approaches concerned are tried. No regulatory scheme is free from difficulties and even where major areas of potential weakness are identified, there may still be progress to be made in comparison with the status quo. A mixed approach of conventional restrictive rules to protect the most significant features and a more flexible, and even experimental, approach beyond that may offer real benefits, and opportunities for learning, even though aspects of the newer schemes fall short of perfection. Indeed, it may be only through such experimentation that effective and robust designs can be found. The issues discussed here are things to think about, not insuperable objections to progress.

2.1 UNCERTAINTY

Uncertainty permeates all aspects of this topic at different levels. The first is the uncertainty in assessing the success or failure of past or new conservation efforts. As Miteva, Pattanayak and Ferraro have shown, it is difficult to demonstrate the effectiveness of any conservation measures.[2] There are many influences other than conservation policy that affect biodiversity and policy may inevitably take a long time to show results.[3] Identifying the results of particular actions is further hampered by the

---

[2] Daniela Miteva, Subhrendu Pattanayak and Paul Ferraro, 'Evaluation of biodiversity policy instruments: what works and what doesn't?' (2012) 28 *Oxford Review of Economic Policy* 69 (published in revised form as 'Do biodiversity policies work? The case for *Conservation Evaluation 2.0*' in Dieter Helm and Cameron Hepburn (eds), *Nature in Balance: The Economics of Biodiversity* (OUP, 2014)).

[3] See e.g., work on the slow recovery and continuing human threats to the blue whale population in the eastern Pacific after hunting ceased; Cole Monnahan and Trevor Branch, 'Do ship strikes threaten the recovery of endangered eastern North Pacific blue whales?' (2015) 31 *Marine Mammal Science* 279.

fact that there are inherent differences between the specific areas chosen to benefit from protective measures and those excluded, whilst the presence of initiatives in any one location or one area of policy will also have some effect on what happens in others. For all these reasons, direct comparisons between locations with and without, or before and after, the implementation of particular conservation mechanisms are unlikely to establish precisely the real effectiveness of the intervention.[4]

Many of the mechanisms discussed here rely on the enhancement, restoration or creation of habitat to balance losses that have been or will be suffered, yet there is uncertainty over whether any such projects can or will achieve their biodiversity objectives.[5] In some cases, such as mature woodland, it is obvious that even in the most successful of projects it will take decades for the replacement of the most obvious elements of the habitat to become established,[6] whilst it remains uncertain how far the rich communities at microbial levels can ever be recreated. Even in the case of habitats that are comparatively quick to establish, such as coastal wetlands, research suggests that the new habitat does not have equivalent biological characteristics to natural sites.[7] Moreover, the success of any project is not guaranteed,[8] with a major commitment required to have a chance of positive outcomes. One successful project in Australia, for example, showed that achieving success called for a much greater area to be committed to habitat creation than the area being lost,

---

[4] As an example of the difficulty of assessing why conservation measures are not producing the hoped for improvements, see John Calladine and others, 'Conservation management of moorland: a case study of the effectiveness of a combined suite of management prescriptions which aim to enhance breeding bird populations' (2014) 61 *Bird Study* 56.

[5] Martine Maron and others, 'Faustian bargains? Restoration realities in the context of biodiversity offset policies' (2012) 155 *Biological Conservation* 141.

[6] Time-lags are an even bigger issue in cooler areas (at higher latitudes or on higher land or at greater depth in the oceans) where growth is slower.

[7] Hannah Mossman, Anthony Davy and Alastair Grant, 'Does managed coastal realignment create saltmarshes with "equivalent biological characteristics" to natural reference sites?' (2012) 49 *Journal of Applied Ecology* 1446.

[8] A study of projects in North Carolina found a success rate of only 74 per cent in wetland schemes, even without using too exacting a standard to judge success; Tammy Hill and others, 'Compensatory stream and wetland mitigation in North Carolina: an evaluation of regulatory success' (2013) 51 *Environmental Management* 1077.

for robust monitoring of progress and for a willingness to invest more resources if the monitoring showed this to be necessary to maintain progress.[9]

Two further layers of uncertainty are added by the dynamic nature of the wider human and natural environment. Policy interventions are never certain in their outcomes and that is all the more so when the intervention includes market-based initiatives where a major feature is the lack of constant detailed regulation. In the environmental field, the volatile and low prices and the surplus of credits in the EU Emissions Trading Scheme has shown the difficulty of guaranteeing that a market-based approach will achieve its objectives,[10] whilst the importance of wider factors is shown by the fact that recent progress towards emission reduction targets can be attributed largely to the recession following the financial crash of 2007–2008 rather than to specific climate policies.[11] Incentives may prove to be more or less successful than expected and stimulate responses which were unforeseen and potentially harmful, as seen in relation to the promotion of biofuels and the belated realisation of the risks that expanded production may pose to biodiversity.[12]

In relation to the physical environment, there are also uncertainties. However well-planned and soundly based an initiative such as a habitat project may be, it can be completely upset by external factors, such as the incidence of disease, invasive non-native species or extreme weather at a crucial point. Such uncertainty is all the greater at a time of climate change when weather patterns are likely to become less stable and changing temperatures are already having an effect on population distributions. Indeed, the fundamental approach that underlies the whole network of protected areas, ensuring the conservation of good habitat in a particular place, is potentially undermined by the scale of climate changes, which means that what are currently suitable locations may in

---

[9] Evan Pickett and others, 'Achieving no net loss in habitat offset of a threatened frog required high offset ratio and intensive monitoring' (2013) 157 *Biological Conservation* 156.

[10] European Commission, *The State of the European Carbon Market in 2012*, COM(2012)652 final.

[11] Climate Change Committee (UK), *Meeting Carbon Budgets: 2014 Progress Report to Parliament* (2014) 55.

[12] A. Webb and D. Coates, *Biofuels and Biodiversity*, Technical Series No. 65 (Secretariat of the Convention on Biological Diversity, 2012).

future no longer provide the conditions that particular species need to thrive. As summed up by an IUCN report looking at habitats in protected areas:[13]

> Carefully planned and managed ecological restoration of protected areas … is a challenging and complex process … Techniques and approaches are still developing and at the same time, unprecedented levels of uncertainty (e.g., associated with changing climate, invasive species and habitat degradation) mean that despite even the best efforts some restoration will not be successful. Global uncertainty, and uncertainty about the degree to which ecological restoration efforts will succeed, impose special responsibilities on those engaged in protected area restoration to act intelligently, resolutely and with humility.

Despite the uncertainties of the background and over the outcomes that may be achieved, any legal approach to biodiversity conservation has to engage with clearly defined rights and relationships. When the approach utilises any elements of market-based instruments or valuation, certain key features need to be established, and fundamental design issues resolved. Many of these have been thoroughly examined in wider contexts, both economic and environmental,[14] but the special aspects of biodiversity must be kept in mind. The various dimensions of the task are interdependent, but a first issue is to identify the 'units of trade', what it is that is being traded or valued (and who owns them).[15] A second is to consider how far these units can be seen as equivalent, so as to enable them to be exchanged. A third is to set the baseline and standards so that one can identify whether the overall aim of any scheme, typically that of ensuring no net loss to biodiversity, has been achieved. Each of these is considered in turn.

2.2 UNITS OF TRADE

In a standard commercial setting there is little difficulty in identifying what is being traded. Goods are either physical objects, or where the market is extended to incorporeal goods, their definition is clearly delineated. Rather than relying on direct exchange and barter, trading is

---

[13] Karen Keenleyside and others, *Ecological Restoration for Protected Areas: Principles, Guidelines and Best Practices* (IUCN, 2012) 5.

[14] James Salzman and J.B. Ruhl, 'Currencies and commodification of environmental law' (2000) 53 *Stanford Law Review* 607.

[15] See section 2.9 below.

then eased by the adoption of a common currency, money, to allow for a wide range of deals that convert goods of one sort into a different form of goods. For biodiversity and ecosystem services there is no such commercial market. Land has a monetary value that will reflect its value for various purposes, which may or may not include providing habitat, but it is the land and its potential that are looked at, not the ecological benefits it is providing. There are some products of 'natural' environments which have a financial value, but usually this is rooted in realising that value by harvesting, which may be deeply destructive to the biodiversity value of the land beforehand. Thus, in a purely commercial setting, the value of standing trees is based on their value as timber when felled, not on their contribution to a healthy woodland ecosystem. The value of a wetland rests on its potential to be converted to a financially profitable use, often by drainage so as to produce crops or support buildings. Ecosystem services are simply left out of all financial calculations in all but the most unusual circumstances.[16]

To create a market or the basis for payments, something that can be traded or paid for must be established in a way which gives it a clear definition and widespread recognition. This is what has happened in the development of markets for greenhouse gas emissions where emissions of various gases are recognised as the basis of a trading currency. A common currency based on carbon dioxide equivalents[17] provides the basis for deals that rely either on a direct trade-off or on the purchase and sale in tonnes of carbon dioxide equivalents of recognised allowances[18] or credits[19] using money as a further currency to enable trade. In this way

---

16 Such as when land is bought to preserve the view from another site or ensure a water supply.

17 Different gases have different greenhouse effects, some considerably greater than carbon dioxide, e.g. over 100 years, methane has over 20 times the impact of carbon dioxide. A simple introduction to the carbon markets and the equivalence of different gases can be found in the FAQ pages on the United Nations Framework Convention on Climate Change; UNFCCC, 'Frequently Asked Questions', https://unfccc.int/ghg_data/online_help/frequently_asked_questions/items/3826.php#11.

18 As under the EU Emissions Trading Scheme; see generally European Commission Climate Action, *The EU Emissions Trading System (EU ETS)* (25 November 2015), available at http://ec.europa.eu/clima/policies/ets/index_en.htm.

19 As in the certified emissions reductions which are at the heart of the Clean Development Mechanism for the Kyoto Protocol; access to all the legal and technical documents and guidance can be obtained through the UN Framework Convention on Climate Change website; UNFCCC, 'What is the clean development mechanism?', http://cdm.unfccc.int/about/index.html.

emissions at one place and time are exchanged for emissions, or their suppression, at another place and/or time.

Greenhouse gases, though, are simple compared with biodiversity, since there is one thing to be measured and standard conversion rates between different gases can be determined.[20] In contrast, so many different things combine to provide a healthy ecosystem that measuring the biodiversity value of a site is not straightforward, quite apart from the difficulty of formulating an agreed valuation method, as discussed in the next section. Habitat may be seen as being important because of the presence of particular species, or because of the wider ecological communities present. Moreover, the same elements may be of more or less value depending on their location, geographically and in relation to other areas of similar or contrasting habitat. Such complexity does not provide the basis for simple trading or recognition of value.

The complexity does, however, provide a strong incentive to explore the establishment of some form of currency in the form of 'biodiversity credits'. If the losses or gains that are arising at a site are once converted into such credits, trading can then be undertaken on the basis of these, rather than at every stage having to explore individual circumstances and carry out direct comparisons between sites. These credits can be based on a more or less sophisticated calculation,[21] looking at variables such as the area of land involved, the condition of the land, the rarity of the habitat or species involved and the time-scales for and difficulty of any habitat enhancement or creation. Establishing such credits satisfies the legal desire to have a clear and definite structure for exchange but clearly involves what may be controversial decisions in the design of the scheme, determining how the many variables are to be taken into account and what weight they are to be given, as well as difficulties in applying such assessments in practice.

A system of biodiversity credits will tend to take a holistic approach to an area of land, but for ecosystem services a more fragmented approach is possible. This has advantages in that it is possible to look at individual aspects of the ecosystem on their own, isolating specific elements in a way that simplifies the calculation that has to be done; for example, considering separately the benefit that a water meadow provides as a buffer offering flood protection to land downstream. The same land will

---

[20] A further difference is that with greenhouse gases, the focus of trade is the *absence* of an emission, and the emphasis is not adding anything in the physical world, whereas biodiversity trading requires the achievement of tangible gains in the real world.

[21] For fuller discussion, see Chapter 4, section 4.3.

also be offering further ecosystem services and there is then the question of whether these can be sold separately or can or should be bundled together, as considered in more detail later.[22]

2.3 EXCHANGEABILITY AND EQUIVALENCE

In all areas of life we have to compare and decide between very different things in weighing the pros and cons of any action and in deciding what we are going to do.[23] A feature of markets is the ease with which parties can make comparative valuations, not only directly (these apples against those apples) but also through conversion into a shared currency which allows disparate things to be given value that can be discussed and exchanged on a shared basis (£5-worth of apples against £5-worth of oranges), even though individuals may vary in how they judge the value as they convert items into that currency. Any currency depends on the notion of fungibility, that any individual unit is the same as any other and can be freely substituted for it. One pound, dollar or euro is the same as any other pound, dollar or euro, and one tonne of carbon dioxide equivalent is the same as any other. These units are exchangeable. With biodiversity, though, the position is rather different. Even if deemed to be of equivalent value (as explored in the next section), one area of wetland is not the same as an area of sand dunes or an area of woodland. They support different ecological communities and contribute different eco-system services. Thus, an offsetting scheme that allows woodland that is being lost to be compensated for by providing new coastal saltmarsh, does nothing for the squirrels whose habitat is lost when the trees are felled. Indeed, even sites with apparently the same habitat type will not be exactly the same, each having its own microclimate and other physical conditions affecting the range and balance of species present and how the site is used.

In exploring this issue, Salzman and Ruhl identify three forms of non-fungibility in environmental trading schemes: of space, of type and of time.[24] In biodiversity terms, non-fungibility of space is a consequence of the locational issues discussed below;[25] even two sites of the same size

---

22 See Chapter 3, section 3.2.

23 This includes deciding which sites and species are to be designated for special attention under a more traditional approach to conservation laws.

24 Salzman and Ruhl, 'Currencies and commodification of environmental law' (n. 14) 625–30.

25 See section 2.7 below.

with the same habitat will have different consequences for biodiversity depending on their location and connections with or isolation from other sites. For example, areas of habitat that provide staging posts in migratory journeys cannot be substituted by sites elsewhere which cannot fulfil this vital function, even though by themselves those alternatives offer much larger and richer sites for the species which are resident in the areas visited briefly by the migrating species. Non-fungibility of type is obvious, since replacing one sort of habitat with another clearly causes gains and losses to quite different species; a hectare of marsh is not the same as a hectare of grassland is not the same as a hectare of forest is not the same as a hectare of coastal sand dunes. Non-fungibility in time arises because of the time-scales in (re-)creating or restoring habitat and the fact that habitat at different stages of maturity has different characteristics. A newly planted woodland today is not a direct substitute for mature woodland that is being felled, and it may take decades before the biodiversity value that is being lost is replaced. In all these ways, elements of biodiversity are not exchangeable in the ways that a currency should be.

There are several possible responses to this challenge.[26] One is to restrict the scope of any offsetting or compensatory arrangements to like-for-like substitution. An offset might be accepted only if it provides habitat the same as the one being lost.[27] This serves to preserve ecological integrity, but is obviously a major constraint on how schemes can operate, undermining the flexibility which is a supposed benefit of a new approach. Without completely preventing substitutions of different sorts of habitat, the market approach can be followed through by imposing premiums or discounts in the valuation so as to encourage like-for-like deals, or at least secure substantial additional gains when there are departures from that approach. A further possibility is to allow exchanges to be made but to provide oversight mechanisms that can identify and intervene if undesirable features start to emerge across the system as a whole, such as disproportionate loss of one habitat (for example, because its location or nature offers the best opportunity for alternative uses), a predominance of certain sorts of replacement habitat

---

[26] These are explored more fully in Chapter 4 in the context of biodiversity offsetting.

[27] This, of course, assumes that truly equivalent habitat can be created (see section 2.1 above) and ignores the argument that a restored or recreated habitat by definition has less value in philosophical terms than an authentic one, as explored in Robert Elliot, 'Faking nature' (1982) 25 *Inquiry* 81 (see Chapter 8, section 8.2).

being provided (for example, because they are easier or cheaper to establish), or a large-scale transfer of habitat sites from areas of development pressure to other regions in the country (where land may be cheaper and more available). The time lag before such effects become obvious will pose a difficulty, as may determining what sort of steps to take to attempt to restore the position, or at least prevent it becoming worse. Nevertheless, some form of system-level oversight allows flexibility to be maintained at the individual level but enables some response if there are unacceptable consequences when the overall effect is considered.

2.4 VALUATION AND 'NO NET LOSS'

Any trading of or payment for biodiversity requires a value to be set on what is being dealt with, and this is a further challenge to overcome. Setting aside for now the ethical issues over whether it is ever acceptable to think of biodiversity in monetary or other terms that allow trading (these are considered further in Chapter 8), a basis for determining the value of what is being exchanged or paid for must be established. Valuation even in monetary units does not necessarily entail seeing elements of nature as commodities,[28] but can be the bedrock of any formalised comparison of different uses of the world around us:

> We want to make clear that expressing the value of ecosystem services in monetary units does not mean that they should be treated as private commodities that can be traded in private markets. Many ecosystem services are public goods or the product of common assets that cannot (or should not) be privatized ... Their value in monetary units is an estimate of their benefits to society expressed in units that communicate with a broad audience. This can help to raise awareness of the importance of ecosystem services to society and serve as a powerful and essential communication tool to inform better, more balanced decisions regarding trade-offs with policies that enhance GDP but damage ecosystem services.[29]

Finding the value that can be used in trade-offs is not straightforward. Although land does have a monetary value, this will predominantly reflect its capacity to produce financial returns, not to support biodiversity, so that an area of woodland will be valued for the timber it can produce, not the habitat it provides. The challenge is therefore to find a

---

28 See Chapter 1, sections 1.6 and 1.7.

29 Robert Constanza and others, 'Changes in the global value of ecosystem services' (2014) 26 *Global Environmental Change* 152, 157.

way of reflecting the value of the trees as they stand and the ecosystem benefits and services that they provide, not just their value on existing commercial markets once they are cut down.

A lot of work has been done on ways to capture a value for biodiversity and ecosystem services.[30] Surveys of this work, such as by Atkinson, Bateman and Mourato,[31] summarise the main assessment methods: adjusted market prices (based on market prices for products, adjusted for distortions such as subsidies and taxes); production function methods (based on treating ecosystem services as an input in a production process); revealed preference methods (based on actual spending patterns and the impact of ecosystem features on prices); and stated preference methods (based on surveys eliciting willingness to pay), and explore further ways of approaching value such as health values and non-use values.[32] Such methods produce results which attract more or less confidence, depending both on technical grounds and on one's initial disposition to accept the enterprise,[33] but undoubtedly they can be difficult and costly to apply and they struggle to capture the full complexity of the real world. In any event, the fact that accounting methodologies can be found should not obscure the need to continue to question what is being done.[34]

Some metric for trade or valuation must be used but stopping short of a conversion into monetary terms may avoid one of the more controversial steps in the valuation process. Indeed, even creating a monetary value for elements of biodiversity for the sake of comparisons does not necessarily entail viewing them as commodities which have value only in

---

[30] E.g. the work of The Economics of Ecosystems and Biodiversity (TEEB) initiative, www.teebweb.org and earlier the OECD, *Handbook of Biodiversity Valuation: A Guide for Policy Makers* (OECD, 2002).

[31] Giles Atkinson, Ian Bateman and Susan Mourato, 'Valuing ecosystem services and biodiversity' in Helm and Hepburn, *Nature in Balance* (n. 2).

[32] *Ibid.* 108–16.

[33] See e.g., the work of Constanza and others attempting to value global ecosystem services and natural capital and the criticisms it has attracted: Robert Constanza and others, 'The value of the world's ecosystem services and natural capital' (1997) 387 *Nature* 253; Robert Constanza and others, 'Changes in the global value of ecosystem services' (2014) 26 *Global Environmental Change* 152; Matthias Schröter and others, 'Ecosystem services as a contested concept: a synthesis of critique and counter-arguments' (2014) 7 *Conservation Letters* 514.

[34] Questioning that should involve many perspectives and disciplines, as shown by Helen Tregidga, 'Biodiversity offsetting: problematisation of an emerging governance regime' (2013) 26 *Accounting, Auditing and Accountability Journal* 806.

financial terms; goods that have a clear commercial value are often treasured (or disdained) for other reasons. In any event, it must be recognised that a degree of valuation is inherent in the more traditional approaches to conservation law, since reaching decisions on whether to designate a site or species as attracting special legal protection inevitably involves judging the comparative worth of different candidates.

Central to whatever methodology is used, the factors to be taken into account must be determined. The English pilot of biodiversity offsetting[35] opted for a comparatively simple approach intended to be applied without the need for expert assistance, taking into account the area of land concerned; the distinctiveness of the habitat being affected or provided; the existing condition of and the difficulty of restoring or recreating the habitat in question; the fit with wider strategies for biodiversity; and the length of time before any replacement habitat matures. This captures many of the most significant variables, but of course leaves vast scope for argument over how acceptable the detailed methodology is and how the different elements should be weighted in any calculation. In this case a particular issue is the preference for a simple approach arguably failing to reflect the true complexity of the natural world by using short-cuts, such as reducing the issue of distinctiveness of habitat to a three-point scale based on lists provided in accompanying guidance.[36] Moreover, this valuation scheme by itself does not capture all the significant aspects in assessing the worth of the sites being considered on both sides of an offsetting arrangement, such as the value to local residents of access to natural sites and the impacts of substituting one habitat with another, however healthy, a long distance away.[37]

Valuation is important not just between sites but for assessing whether there has been 'no net loss'. No net loss is an attractive concept, seeking to ensure that whatever arrangements are made to deal with development and changes in land use there is no further loss of biodiversity. It is a concept that is increasingly being adopted in various policy documents,[38] despite

---

35 Department for Environment and Rural Affairs, *Biodiversity Offsetting Pilots: Guidance for Developers* and *Guidance for Offset Providers* (2012); for further discussion see Chapter 4, section 4.3.5.

36 Indeed, the approach was described as 'overly simplistic' by the House of Commons Environmental Audit Committee, *Biodiversity Offsetting*, HC 750 (2013–14) 9.

37 *Ibid.* 11–12.

38 The EU's No Net Loss Initiative stemming from the policy commitment in the *EU Biodiversity Strategy to 2020*, COM/2011/0244 final, Action 7(b), is

debates over its meaning.[39] Nevertheless it must be remembered that working on the basis of a 'net' loss or gain for biodiversity requires an acceptance of the ideas that elements of biodiversity can be valued in an accurate way and that they are, to some extent at least, interchangeable, ideas which are not in themselves uncontroversial.[40] Moreover, any assessment of net loss requires a baseline against which changes can be measured, and in most circumstances this simply does not exist. A single survey done in the immediate run-up to a development project can hardly hope to capture the full richness of the biodiversity of a site. Predictions of impact are uncertain, both for the losses on one site and the gains from any compensatory measures, and issues remain over the time-scale across which assessment should be made.[41] Deciding whether biodiversity has gained or lost is not easy, since there will be winners and losers from any changes to the environment. A further issue is how far the declining state of biodiversity is taken into account in setting the baseline against which gains are measured, such that any actions which slow the decline that will flow from 'business as usual' might be regarded as being positive, even though losses are still occurring.[42]

2.5 ADDITIONALITY

The issue of additionality must also be considered where any scheme seeks to reward or recognise the biodiversity benefits of activities that are being undertaken, such as the way in which land is used or managed.[43] If the scheme is to show that it is delivering worthwhile gains, it must be shown that it is achieving something beyond what would have happened

one example; see European Commission Environment, *No Net Loss* (19 November 2015), available at http://ec.europa.eu/environment/nature/biodiversity/nnl/index_en.htm.

[39] Toby Gardner and others, 'Biodiversity offsets and the challenge of achieving no net loss' (2013) 27 *Conservation Biology* 1254.

[40] Similar concerns affect alternative formulations, such as that in New South Wales, where biodiversity certification allows development to be approved in circumstances where, once bio-banking arrangements are taken into account, the project 'will improve or maintain biodiversity values'; Threatened Species Conservation Act 1995, s. 127O.

[41] See section 2.6 below.

[42] Ascelin Gordon and others, 'Perverse incentives risk undermining biodiversity offset policies' (2015) 52 *Journal of Applied Ecology* 532, 533–4.

[43] Karen Bennett, 'Additionality: the next steps for ecosystem services markets' (2010) 20 *Duke Environmental Law and Policy* 417.

in any case, whilst those providing the resources to pay for the scheme will want to be reassured that they are getting something in return and that nobody is getting paid more than once for the same thing. Assessing additionality will be difficult because of the wider issues of measurement and valuation (discussed above), especially in long-term projects, and also because of the interaction with other policy measures. Even in the more established carbon offset markets it is not always clear whether a claimed offset is in fact achieving something other than 'business as usual'. The conclusion of a recent study of such markets was that 'it remains technically challenging to gauge whether many offsets are genuine reductions'.[44] Where direct controls exist to prevent harm to certain irreplaceable or otherwise exceptionally valuable habitats, simply complying with those laws is achieving no real gain that should be paid for[45] or should be claimed for offsetting against other harms. Thus the bio-banking scheme in New South Wales explicitly states that for any project to be accepted as a recognised offset, it must achieve more than is already required under any of a series of other conservation measures.[46] A discounting mechanism is applied where there is a less complete overlap between what is required by specific land management obligations and what is required to achieve the overall gains sought.[47]

The situation where there is a direct overlap between other legal requirements and a project proposed as an offset is fairly straightforward compared to others. Where there is no such direct correlation, many factors which influence how land is managed will have to be considered in determining whether there is a biodiversity gain over and above what might be expected by 'business as usual'. Moreover these factors are likely to change over time, as shown by the transformation of agri-environment requirements within EU agricultural policy from being a limited option within special schemes to forming a basic requirement of any financial support.[48] What would have been expected from 'business

---

44 Christopher Arup and Hao Zhang, 'Lessons from regulating carbon markets' (2015) 4 *Transnational Environmental Law* 69, 100.

45 We are not entitled to collect an annual fee for every year we manage to get through without murdering anyone.

46 Threatened Species Conservation (Biodiversity Banking) Regulations 2008, reg. 4.

47 Department of Environment, Climate Change and Water NSW, *Additionality and its Effect on Credit Creation* (Department of Environment, Climate Change and Water NSW, 2010).

48 Whereas the first EU schemes were options which Member States might choose to implement (Council Regulation (EEC) 797/85 of 12 March 1985 on

as usual' 30 years ago is very different from what could be expected today, and this evolution is a challenge for mechanisms attempting to provide for long-term habitat protection or enhancement.

The comparison with what would happen without the intervention in question is central to the assessment of additionality in relation to the Clean Development Mechanism (CDM) under the Kyoto Protocol, where considerable work has been done to define the concept and investigate whether it is being achieved.[49] 'A CDM project activity is additional if anthropogenic emissions of greenhouse gases by sources are reduced below those that would have occurred in the absence of the registered CDM project activity',[50] and specific further definitions and guidance are given for different forms of project.[51] Each project must specify how it is additional and this is checked as part of the verification process,[52] yet serious doubts have been raised as to the methodologies used and the effectiveness of the practice in assessing additionality.[53] Similar measures

---

improving the efficiency of agricultural structures, [1985] OJ L93/1, art. 19), now compliance with environmental standards is a requirement of any payments (Council Regulation (EC) 1782/2003 of 29 September 2003 establishing common rules for direct support schemes under the common agricultural policy and establishing certain support schemes for farmers and amending Regulations (EEC) 2019/93, (EC) 1452/2001, (EC) 1453/2001, (EC) 1454/2001, (EC) 1868/94, (EC) 1251/1999, (EC) 1254/1999, (EC) 1673/2000, (EEC) 2358/71 and (EC) 2529/2001, [2003] OJ L27/1, arts 3–5).

[49] Benito Müller, *Additionality in the Clean Development Mechanism*, Oxford Institute for Energy Studies EV 44 (2009).

[50] Framework Convention on Climate Change, Meeting of the Parties (2005) 3/CMP.1 Annex para. 43; FCCC/KP/CMP/2005/8/Add.1.

[51] The definitions and procedures for different forms of project can be conveniently found in the *CDM Rulebook*, a database prepared by Baker & McKenzie (look under 'Additionality' in the A-Z index): http://cdmrulebook.org/. This source also provides useful links to the original documents.

[52] *Clean Development Mechanism: Validation and Verification Manual (Version 01.2)*, FCCC CDM-EB 55, Annex 1, especially para. 94.

[53] Axel Michaelowa, 'Interpreting the additionality of CDM projects: changes in additionality definition and regulatory practices over time' (248–71) and Christiana Figueras and Charlotte Streck, 'A post-2012 vision for the Clean Development Mechanism' (564), both in David Freestone and Charlotte Streck (eds), *Legal Aspects of Carbon Trading: Kyoto, Copenhagen and Beyond* (OUP, 2009); Lambert Schneider, *Is the CDM Fulfilling its Environmental and Sustainable Development Objectives? An Evaluation of the CDM and Options for Improvement* (Report prepared for WWF, Öko-Instutut eV, 2007) 27–45, 58–61; on the similar issues arising under the Joint Implementation Scheme, see Anja Kollmuss, Lambert Schneider and Vladyslav Zhezherin, *Has Joint Implementation Reduced GHG Emissions? Lessons Learned for the Design of Carbon*

could be instituted for biodiversity projects, but any project-specific approach may add significant complexity and cost, whilst alternatives bring different potential and problems.[54]

A further issue for biodiversity applications is how far simply maintaining the status quo can be regarded as providing something additional. Securing the long-term protection of an area of wetland against development of any sort may be seen as a benefit, but if that area sits in a remote area and would require expensive drainage works before any construction or change of use could be contemplated, development is not likely in the foreseeable future and therefore there is not really any additional benefit that has been gained. On the other hand, ensuring the conservation of coastal areas near a city that are under very real development pressure may represent something that is very much worth obtaining for the benefit of biodiversity. The same question has arisen in various contexts where money is paid in return for steps that prevent damaging activities taking place. Examples include the tax breaks obtained in exchange for some conservation easements in the United States where there seemed no real likelihood of damaging land use changes actually taking place,[55] and payments for maintaining forest cover in Costa Rica, where the difficulty of assessing additionality and limited achievements are reported.[56] Similarly, at a time when in Great Britain compensation was available for the profit foregone by refraining from operations that would damage a protected area (Site of Special Scientific Interest), there was doubt over whether some development proposals were ever seriously intended or had been put forward merely to attract the compensation for the theoretical loss of profit.[57]

A different perspective is to place less emphasis on additionality and to take the view that the whole idea of payment for ecosystem services or for providing an offset is to reward landowners for leaving land in a

---

Market Mechanisms, Stockholm Environment Institute Working Paper No. 2015-07 (2015).

[54] Bennett, 'Additionality' (n. 43) critically examines project specific approaches, standardised additionality assessments, discounting ratios and probability assessment.

[55] Daniel I. Halperin, 'Incentives for conservation easements: the charitable deduction or a better way' (2011) 74 *Law and Contemporary Problems* 29, 43. The payments in such cases may be direct or through tax breaks which have been identified as the main incentive for such easements to be granted; *ibid.*

[56] Amy E. Daniels, and others, 'Understanding the impacts of Costa Rica's PES: are we asking the right questions?' (2010) 69 *Ecological Economics* 2116.

[57] Scottish Office, *People and Nature: A New Approach to SSSI Designations in Scotland* (1998) 28.

quasi-natural state that is providing services which are not currently being paid for. Therefore, even when just maintaining the status quo, landowners are not being rewarded for 'doing nothing' but for continuing to supply benefits of value to wider society and which have in the past gone unrewarded. Payments therefore serve to correct a past imbalance and to reduce the incentive to convert the land to other uses producing more conventional income streams. Nevertheless, until there is widespread acceptance of the currently under-appreciated benefits provided by ecosystem services, it is likely that for conservation schemes to have legitimacy, the steps supporting biodiversity that market mechanisms recognise must produce a true gain for biodiversity or else providers will be seen as receiving benefits (whether payment or offset credits) under essentially false pretences.

A further aspect of additionality can arise if schemes are successful. It is common to offer financial support to farmers and others who suffer losses when rare species are being supported or reintroduced to an area, for example, to compensate for losses caused by carnivores taking livestock.[58] If such schemes succeed, though, and the species thrive, not only do the costs increase but the justification for payments as supporting something special or additional declines. At what stage does the enhanced biodiversity become the 'new normal' so that there is no additionality to justify intervention? This may be a real concern to far-sighted landowners who see the end result of their co-operation in special conservation schemes as their being left with a changed environment where they face additional burdens which will no longer be shared with the conservation bodies who have encouraged those changes.[59] The conservation focus, and budgets, move on, but the landowners look back and see themselves on their own as having acquired additional costs to be borne into the future.

---

[58] Such schemes are widespread, covering species such as wolves, bears and white-tailed eagles but also extend to species causing damage in less dramatic ways, such as grazing by over-wintering geese; e.g. between 1987 and 2009 the Wolf Compensation Trust operated by Defenders of Wildlife paid US$1,368,043 to compensate for livestock killed by wolves in parts of the United States, see www.defenders.org/publications/statistics_on_payments_from_the_defenders_wildlife_foundation_wolf_compensation_trust.pdf.

[59] Hence the development of 'safe harbor' agreements in the United States where landowners participating in a scheme to support the recovery of endangered species are guaranteed that no additional restrictions will be imposed on them; for a summary see US Fish and Wildlife Service, *Safe Harbor Agreements for Private Landowners* (July 2011), available at www.fws.gov/endangered/esa-library/pdf/harborqa.pdf.

2.6 DURATION AND TIMING

Time-scales provide a further challenge to thinking about how biodiversity can be encouraged and managed, affecting both the delivery of biodiversity benefits and the regulatory approach. Any system that takes account of the enhancement or (re-)creation of habitat must cope with the fact that it may be decades before a healthy, mature ecosystem can be realised. This is at its most obvious with woodland, where it clearly takes decades for trees to reach any appreciable size and may take several centuries before there is a varied woodland community, with a mix of young, mature, senescent and dead trees that can provide the range of ecological niches required to fulfil the biodiversity potential of the site. Some biodiversity benefits may be achieved at an early stage, but most must wait until a degree of maturity has been reached; indeed, different ecological communities may benefit at different times as an area matures. Even with habitats that can mature much faster,[60] there will still be some time-lag between starting a conservation project and realising its full benefits, so that there will be a temporal gap between the losses and gains in an offsetting deal.

There are also temporal aspects in relation to ecosystem services, which again are delivered over a considerable time-scale. It is conceivable that some ecosystem services could be restricted to a single occasion, for example, allowing land to be used to provide an alternative watercourse in response to a wholly extraordinary blockage to the normal channel, but usually such services are not a one-off event and it is their provision over an extended period that is of value to society. Moreover, most services depend on ecological communities or features which cannot be turned on and off easily, but rely on the continuing long-term health of the ecosystem which takes time to establish.

Any scheme that recognises biodiversity benefits or ecosystem services must therefore take account of the timing issue. If biodiversity is to be secured, there should be long-term arrangements in place, and if offsets are to be recognised as compensating for biodiversity losses, then since the loss is permanent it can be argued that the balancing benefit should

---

[60] Habitats such as meadows may appear to regenerate more quickly but even then it may be decades before the functional traits of the plant community are replicated and over 150 years for all the appropriate species to establish themselves; Ben A. Woodcock, Alison W. McDonald and Richard F. Pywell, 'Can long-term floodplain meadow recreation replicate species composition and functional characteristics of target grasslands?' (2011) 48 *Journal of Applied Ecology* 1070.

be as well.[61] The long time-scales set challenges in relation to the establishment of monitoring and enforcement arrangements that can still be relied on many years in the future. There are also practical problems since few individuals or institutions are willing to commit themselves to perpetual, or even fairly long-term, legal obligations. Landowners may be happy to commit their land for a decade or so, but are unwilling to tie the hands of the next generation who will be faced with producing an income from the land in very different circumstances. Thus most existing schemes, such as those deriving from EU agri-environment measures, are based on fairly short time-spans.[62] An exception to this is provided by conservation easements in the United States which do entail a commitment in perpetuity. This feature, though, is driven largely by the requirements that have to be met for the creation of an easement to be treated as a charitable gift which attracts substantial tax breaks and provides a key incentive to enter such agreements.[63]

A perpetual, or very long-term, obligation raises its own problems, not just because future generations may not be in sympathy with its objectives and therefore more willing to challenge, or ignore, its terms.[64] An obligation intended to protect a particular species at one location may make no sense if climate change, disease or other causes mean that in future that species is no longer present there. Therefore a balance must be sought between sufficiently secure long-term commitments and the flexibility to cope with changing circumstances. The drafting of obligations must therefore be carefully considered, with the potential for using a mix of longer and shorter term provisions in order to secure the main objective over the long run whilst allowing flexibility in specifying the shorter-term actions to be taken to achieve that. Such an approach is in

---

61 See Chapter 4, section 4.3.

62 E.g. the financial support for the management or restoration of lowland raised bogs in Scotland under the Rural Priorities within the Scottish Rural Development Programme is based on merely a five-year commitment; Scottish Government, *Management/Restoration of Lowland Raised Bogs* (2015), available at www.scotland.gov.uk/Topics/farmingrural/SRDP/RuralPriorities/Options/LowlandRaisedBogs.

63 See Chapter 5.

64 Anecdotal evidence from the United States suggests a growing number of cases where conservation easements are coming under question from those who succeed the original creator. These heirs or other successors may not share the same conservation motive that led to the creation of the easement and they are less likely to share the sense that there is a 'bargain' to be observed which is likely to affect those who were the direct recipients of the tax breaks enjoyed when the easement was created.

line with an adaptive management approach, creating scope to learn and respond as our knowledge and the factual situation change, but it does create a risk of the flexibility being misused.

Timing is also significant because of the time-lag between starting operations and delivering a biodiversity benefit. This is significant particularly in offsetting and compensatory schemes, where the loss of habitat today is being balanced by gains that will only materialise in the future – or rather, given the uncertainty of most habitat (re-)creation and enhancement, gains that may or may not materialise in the future. This can just be ignored or accepted as a feature inherent in the scheme,[65] or more responsibly it can be taken into account by discounting the value of the gain to reflect the time-lag[66] or by demanding that additional interim gains are provided until the longer-term balance can be achieved.[67]

The time-lag also affects who should be rewarded for biodiversity gains. The person in charge of land when a biodiversity project is started may no longer be responsible for the site when it delivers its benefits. Since many of the costs of major habitat work are front-loaded, falling at the beginning of a project which will only 'produce a return' in years to come, there is a question over who should be given the rewards for delivering the benefits, the person who bore the cost or the one responsible at the time of actual delivery. Resolution of this requires attention to the input/output question discussed in section 2.13 below, and possibly also to ways to spread costs and rewards over a longer time-scale rather than being concentrated where they directly arise.[68]

---

[65] E.g. a crude carbon offset scheme might allow credits today based on a new afforestation project, even though it will be some years before the trees are established and begin any carbon sequestration.

[66] As in the English biodiversity offsetting pilot, where a discounting rate applied to reduce the value of an offset being provided, depending on how long it would take for a site to reach its target condition; DEFRA, *Biodiversity Offsetting Pilots: Guidance for Offset Providers* (DEFRA, 2012); see Chapter 4, section 4.3.5.

[67] As under the EU's Environmental Liability Directive where 'compensatory' remediation is required to compensate for the loss of natural resources or services pending their recovery or the achievement of longer-term 'complementary' remediation; Directive 2004/35/EC of the European Parliament and Council on environmental liability with regard to the prevention and remedying of environmental damage, [2004] OJ L143/56, Annex II.

[68] Cf. the UK Government's 'Green Deal' (closed to new applications in autumn 2015) under which the cost of energy efficiency improvements are repaid over a period of time, tied to the electricity bill for the premises and thus falling on the occupier from time to time rather than being treated as a one-off liability:

A further human aspect of the extended time-scales must be considered as well as the ecological side. The original parties to an agreement will not be around for ever; an original party may want to transfer the obligation or the property to which it is tied, individuals will die, companies may be taken over or become insolvent, public authorities may be reconstituted, merged or abolished and charities or other associations may be dissolved. Clear arrangements must be in place to enable obligations to have continuing force once the original parties are no longer involved. Given the central importance of land management, ensuring that obligations can 'run with the land', binding successors in title to the landowner who enters an agreement, is an obvious step to take. But the other side of the agreement must be considered as well. If a body enters a conservation covenant with a landowner, it should also be clear how the benefit of that obligation can be passed on.[69] Non-governmental organisations change over time, merging and disappearing, whilst government bodies are forever subject to reorganisation and restructuring. Provision should be made to ensure that obligations that are intended to be enduring can be passed on appropriately so that the fate of biodiversity is not dependent on the fact of one particular body continuing in a specific form.

Similar considerations affect the financial side of any arrangement. To the extent that schemes are designed to provide income in the future, whether to pay for conservation management activities or as a reward for delivering ecosystem services, there must be a mechanism to achieve this. This could simply require the identified beneficiaries of a scheme to continue making the payments as and when due. However, especially in the context of development projects under an offsetting scheme, there may well be advantages in substituting an up-front payment to establish a fund which will meet the future liabilities. This reduces the dangers posed by the risk of a beneficiary becoming unwilling or unable to pay in the future, a particular danger where the party has no continuing connection with a project, as is the case with developers whose business is based on completing the initial development and then selling on, rather than taking long-term responsibility for the site. Similar dangers arise with projects such as mines, quarries and waste sites, where there may be permanent impacts on the land but the operators' income flows continue only so long as they are able actively to exploit the site, with particular

---

Green Deal: Energy Saving for Your Home (Gov.UK, 16 October 2015), available at www.gov.uk/green-deal-energy-saving-measures/overview.

[69] See Chapter 5, section 5.2.

challenges in the event of insolvency.[70] Other ways of meeting this risk could be to insist on appropriate insurance, bonds or other guarantees[71] to ensure payment in the event of default.[72]

2.7 LOCATION

Locational issues[73] are important for biodiversity mechanisms in at least four ways. Two biological aspects relate to the suitability and the connectivity of sites, whereas more social factors concern their availability and their relationship with the human population. In all four of these ways there is a big difference between dealing with biodiversity and with the sorts of pollution (acid rain, greenhouse gases) which have been the focus of earlier market-based approaches. Although there will be site-specific issues to consider in pollution cases,[74] these are dealt with largely by other regulatory mechanisms and it is the overall concentration across a large area that is addressed by trading schemes and incentives. For biodiversity, locational issues even at a very small scale are vital.[75]

The biological suitability issue is in some cases straightforward, since it is clear that any offset or habitat enhancement designed to assist a particular species, or community of species, must provide the appropriate conditions for it to thrive. Land may be available in the Sahara Desert, but it cannot be used to provide habitat for polar bears, nor would Arctic tundra be suitable for giraffes. The requirements of many species are

---

70 *Scottish Environment Protection Agency* v. *Joint Liquidators of the Scottish Coal Company Ltd* [2013] CSIH 108.

71 Robert L. Glicksman, 'Regulatory safeguards for accountable ecosystem service markets in wetlands development' (2014) 62 *Kansas Law Review* 943, 956–7.

72 Care must be taken in setting up and drafting such arrangements so as to minimise the difficulties if circumstances do require their activation, as shown by the disputes over the remediation costs for open-cast mines following the insolvency of Scottish Coal: *East Ayrshire Council* v. *Zurich Insurance plc* [2014] CSOH 102; *South Lanarkshire Council* v. *Coface SA* [2016] CSIH 15.

73 Philip Womble and Martin Doyle, 'The geography of trading ecosystem services: a case study of wetland and stream compensatory mitigation markets' (2012) 36 *Harvard Environmental Law Review* 229.

74 Extreme local concentrations of carbon dioxide, even those arising from natural sources, can prove fatal, as shown by the 1,700 deaths at Lake Nyos in Cameroon in 1986.

75 Even microhabitat issues can affect significantly our awareness of the biodiversity in any area; Zia Mehrabi and others, 'The importance of microhabitat for biodiversity sampling' (2014) 9(12) *PLoS ONE* e114015.

much more finely tuned than these extreme examples; indeed, the story of evolution and biodiversity is one of marginal differences creating the niches in which particular species or ecological communities can flourish at the expense of others. Temperature, rainfall, the salinity of waters, the geology of the underlying rocks, the acidity of the soil, the presence or absence of other species from the microbial level upwards and many other variables can be vital to whether a location is suitable for a species to survive. There are therefore only so many places where it is possible to provide particular ecosystem services or to establish an offset to benefit particular species, so that the scope for substitution may be very restricted. Moreover, the benefits of maintaining discrete local populations[76] may also call for sites to be provided in the vicinity of those under threat,[77] rather than proposing that a species with fragmented but widespread distribution be concentrated in one area, however suitable that one location may be.

Connectivity is another key aspect of the value of sites for biodiversity. It is not just what is on a site itself but how it connects with other sites that determines how much a site can contribute to the health of biodiversity in the area.[78] Thinking about this fits in with the ecosystem approach to nature and is something which traditional conservation law is not always good at addressing, with its emphasis on discrete protected sites, too often cut off from other areas and leaving species isolated within a hostile background. Connectivity is recognised in some existing measures, such as the aim of the EU Habitats Directive to provide a 'coherent ecological network' of protected sites[79] and its encouragement of the conservation of features which:

> by virtue of their linear and continuous structure (such as rivers with their banks or the traditional systems for marking field boundaries) or their

---

[76] Both for inherent diversity between populations and to provide greater resilience by avoiding having 'all the eggs in one basket'.

[77] The US wetland mitigation scheme requires any offset to be provided in the same watershed; 33 CFR §332.3(b)(1), 40 CFR §230.93(b)(1).

[78] Kevin R. Crooks and M. Sanjayan (eds), *Connectivity Conservation* (CUP, 2006); Graeme L. Worboys, Wendy L. Francis and Michael Lockwood (eds), *Connectivity Conservation Management: A Global Guide* (Routledge, 2010).

[79] Council Directive 92/43/EEC of 21 May 1992 on the conservation of natural habitats and of wild fauna and flora (EU Habitats Directive), [1992] OJ L206/7, Preamble, para. 10.

> function as stepping stones (such as ponds or small woods), are essential for the migration, dispersal and genetic exchange of wild species.[80]

The potential for such movement and interchange is all the more important at a time of climate change when species may need the scope to move to new areas as their existing homes no longer suit their needs as a result of increased temperatures or changes in weather conditions or sea-level.

Connectivity is thus relevant in assessing sites for biodiversity purposes. Apparently identical sites may have more or less value in providing ecosystem services or as an offset depending on their connections with other habitats. Moreover, sites which may not have a great inherent value by themselves may be valuable as a link between other sites, which would in turn be greatly diminished if this link was lost.[81] Thus, the value to be attributed to a hectare of degraded woodland may vary greatly depending on whether it is adjacent to other woodland in better condition (and so has ready potential to become enhanced), it is standing on its own (where it is vulnerable and less likely to develop into a richer habitat without significant intervention), or although of no great merit or potential in itself it serves to connect other sites of greater interest. Even when sites are in isolation, their value will vary in line with the connections with the surrounding environment: an isolated stand of trees may be doomed never to flourish if its environs are and remain unsympathetic, or may provide a priceless refuge for species, the core from which the whole area can be regenerated, if the surrounding environment becomes more welcoming.

Turning to the human aspects of location, the first point to note is the simple one that even where there are biologically suitable sites that can be used for habitat creation or conservation, they may not be available for that use because of other demands on them. In less densely settled countries where there are large expanses of uncultivated or undeveloped land this may be less of a constraint, but in a densely populated country like the United Kingdom, where almost all the land is devoted to commercial purposes of one sort or another, finding land that can be dedicated for conservation purposes may be a difficulty. The competing demands on land may also be meeting other important social needs, such as food production and housing, so that it is not simply a question of prioritising conservation over private gain. Even where recourse can be

---

[80] *Ibid.* art. 10.

[81] Andrew Bennett, *Linkages in the Landscape: The Role of Corridors and Connectivity in Wildlife Conservation* (IUCN, 2003).

had to compulsory purchase to ensure that land is released as opposed to being kept for other purposes, the process may not be straightforward, as shown by the protracted litigation associated with providing sites to compensate for the loss of protected mud flats when the Cardiff Bay Barrage was built.[82]

Moreover, as that litigation shows, the demand for habitat under an offsetting or similar scheme can have a major impact on financial value. One of the purposes of introducing such schemes is to change the current position and ensure that providing for biodiversity becomes an economically valuable use of land. Nevertheless, when combined with the scarcity of suitable sites and the absence of a wider 'biodiversity market', the demand for a specific offset can have a very substantial and perhaps distorting effect on prices. In the Cardiff case, land that was 'indispensable' to provide the compensatory measures required under EU law was valued by the claimants at £4,500 per acre at agricultural value, at £13,000 per acre as a nature reserve but at £28,000 per acre when its vital role in the development scheme was considered (described at one stage in proceedings as 'ransom value').[83] Wider schemes which create the demand for offsets should in turn promote greater supply and competition and allow prices to settle, but some categories of sites will always be in short supply.

The final aspect of location is how humans interact with their surroundings. When there are proposals to develop a site and it is assessed in terms of biodiversity alone, an area of poor quality scrub or meadow close to lots of houses may be a lot less valuable than a plot of similar habitat in better condition elsewhere. Even an area of the same quality will bring more biodiversity benefits if it is located away from disturbance and connected to areas of richer habitat, where there are gains both in providing a larger area of habitat and in the potential for the poorer area to be enhanced through the spread of species. On this basis, when it comes to offsetting (or any other scheme based on equivalences) it would make sense in ecological terms to develop the first site but to compensate for it by providing an area of habitat where it can bring the most benefits, even though this is a long way from the original location.

Such calculation, however, ignores the human dimension. The first site may not be of great value from an ecological point of view, but losing it may be a great loss to local residents. One of the justifications for

---

82 *Waters* v. *Welsh Development Agency* [2004] UKHL 19, [2004] 1 WLR 1304.

83 *Ibid.* [10]; the court decided that the middle value was the appropriate one to use.

conserving biodiversity is its value for human well-being,[84] and this is undermined if the result of an offsetting or other scheme is in effect to transfer all the sites where people can interact with nature away from where people live. At its extreme, a scheme that bases calculations purely on ecological values might allow an area subject to development pressure to be almost completely built over, whilst 'nature' is sent away to a distant region. The ecological benefits may be greater, but the loss to human well-being should also be taken into account: 'Access to nature is vital'.[85] This is an increasingly significant issue as the link between health and access to nature becomes more prominent in biodiversity policies, for example, in Scotland where providing 'quality greenspace for health and education benefits' is identified as one of the 'Six Big Steps for Nature' through which biodiversity policy is to be delivered.[86] Identifying the need to consider the human dimension does not, of course, show exactly how this should be done. If residents are still to have access to nature locally, how local is that (walking distance, so many minutes' driving distance?) and how far should ecological (such as water catchments) or social (such as local authority) boundaries determine the question?

2.8 COHERENCE

The need for co-ordination and coherence is a further consideration which makes tackling biodiversity matters different from dealing with many other environmental problems. A conservation programme that is based not on co-ordinated activity by state bodies and a network of designated sites, but on the cumulative effect of individual biodiversity offsets, conservation covenants and ecosystem service agreements, offers much more flexibility and scope for individual initiative. This can be a positive feature, but also a negative one. For greenhouse gases it does not matter much which emissions are actually reduced so long as the overall

---

84 'Connecting with nature enriches our lives. Many benefits arise from this across a range of policy areas, including health, wellbeing, education, community development and regeneration. Better integration of environmental and social objectives will undoubtedly secure long-term benefits for people and nature.' Scottish Government, *2020 Challenge for Scotland's Biodiversity: A Strategy for the Conservation and Enhancement of Biodiversity in Scotland* (2013) 33.

85 House of Commons Environmental Audit Committee, *Biodiversity Offsetting*, HC 750 (2013–14) 12.

86 Scottish Government, *Scotland's Biodiversity: A Route Map to 2020* (2015).

concentration of atmospheric carbon dioxide is reduced; methane reductions in New Zealand can be looked on as essentially the same as carbon dioxide reductions in the United States once the appropriate conversion factors have been applied. For biodiversity things are very different, especially in two of the ways discussed above. In the first place, locational issues are vital. Sites may have some value in isolation, but they have much more when seen as part of a wider network. It is significant, for example, that EU biodiversity policy has at its heart the Natura 2000 *network* of protected sites,[87] and that on the occasions when overriding reasons justify projects that damage sites, states then have to 'take all compensatory measures necessary to ensure that the overall coherence of Natura 2000 is protected'.[88] Leaving to the open market the choice of where ecosystem services are protected and which sites are to be lost or enhanced removes the ability to ensure that conservation effort results in a coherent pattern of sites suitable for the needs of a wide variety of species.[89] If an offset policy resulted in development being concentrated in one area and conservation sites in another, this might at least constrict the distribution of species and disrupt ecological communities, and at worst endanger their survival by denying the spread of sites needed at different times of the year. There is no point in conserving the feeding grounds of bats, if there is nowhere suitable for them to roost during the winter.

Such considerations suggest that measures must be put in place to provide a degree of co-ordination and coherence in the overall outcomes of the conservation mechanisms used. National biodiversity strategies, plans and programmes are required under the Convention on Biological Diversity,[90] and they can be used to shape schemes and to guide the various parties involved, but it may be necessary to include incentives and constraints to ensure that the overall outcome of individual agreements is that appropriate results are achieved. 'Central planning' is the

---

[87] European Commission, *Natura 2000 Network* (19 November 2015), available at http://ec.europa.eu/environment/nature/natura2000/index_en.htm.

[88] Council Directive 92/43/EEC of 21 May 1992 on the conservation of natural habitats and of wild flora and fauna, [1992] OJ L206/7, art. 6(4).

[89] Strengthening the monitoring and evaluation at the scheme-wide level and increasing the significance of the strategic location of sites are among the recommendations in the review of the operation of the New South Wales BioBanking scheme; New South Wales Government, *BioBanking Scheme: Statutory Review Report* (2014) 9 and 23.

[90] Convention on Biological Diversity (adopted 5 June 1992, entered into force 29 December 1993) 1760 UNTS 79, Art. 6.

antithesis of a market approach, but an element of this may be needed to ensure an effective spread of conservation effort.[91]

Secondly, the different elements of biodiversity do not share the exchangeability of greenhouse gases. Allowing the market to decide that all of the sand dunes along a coast are to be lost and offset by the creation of woodlands, does not ensure the survival of biodiversity, however carefully calculations have been done to measure the equivalence of the various sites involved. Likewise, concentrating on only some ecosystem services at the expense of others will not guarantee an appropriate spread of effort and a healthy overall environment. Again, a degree of planned coherence, the opposite of the fluidity of the market, is required to ensure that conservation activity actually results in benefits for biodiversity.

2.9 PROPERTY RIGHTS

A greater emphasis on private rights lies at the heart of many of the mechanisms emerging as an alternative to direct regulation in the attempt to further biodiversity.[92] One of the prerequisites for a working market is a recognition of and respect for property rights and entitlements.[93] Schemes based on payment for ecosystem services rest on an identification and legal recognition of who exactly has the right to claim payment and for what exactly they are being paid, while biodiversity offsetting schemes operate on a basis of creating (in some form) biodiversity 'units' which can be held and traded. Such developments focus attention on the question of the ownership of 'nature' in a way that the law has not previously had to deal with.[94]

The traditional legal background recognises private ownership of land (and plants as accretions to the soil) but not of wild creatures, and most ecosystem services are not reflected in legally recognised entitlements. There are no legal rights which entitle landowners to charge for the

---

[91] See Chapter 3, section 3.5.

[92] OECD, *Handbook of Market Creation for Biodiversity: Issues in Implementation* (OECD, 2004) ch. 3.

[93] John McMillan, *Reinventing the Bazaar: A Natural History of Markets* (Norton, 2002) ix.

[94] Walters Nsoh and Colin T. Reid, 'Privatisation of biodiversity: who can sell ecosystem services?'(2013) 25 *Environmental Law and Management* 12; Colin T. Reid and Walters Nsoh, 'Whose ecosystem is it anyway? Private and public rights under new approaches to biodiversity conservation' (2014) 5 *Journal of Human Rights and the Environment* 112.

services of the wild pollinating insects which travel onto their neighbours' land nor which entitle the neighbour to demand these services (although bee-keepers can and do sell the services of the bees they own). The extensive rights of the owner over the land provide a useful starting point for considering how legal relationships fit with new approaches to conservation. Conventional legal structures can be utilised to the extent that conservation goals can be secured simply by restricting the owner's freedom of action, for example, by prohibiting building or other forms of development which will destroy habitat.[95] Even here, though, there are difficulties arising from the Anglo-American tradition that is unwilling to accept enduring restrictions on the use of land except for the benefit of neighbouring property, requiring statutory intervention to enable long-term conservation easements or covenants to be created.[96]

A clearer role for private rights emerges in relation to the trading aspect of biodiversity offsets and payment for ecosystem services. Under payment schemes it must be determined what it is that is being sold and who can sell it. The provision of ecosystem services is directly related to what happens on the land, yet there may be several parties with rights in the land that affect its management and therefore the ecosystem services that the land can deliver. The existing allocation of rights and responsibilities between landowner, tenant and holder of lesser rights (such as rights over minerals or hunting) will make a big difference to who has the power to decide how the land is managed and what services are provided. Decisions by any one of these on whether and how to exercise their legal entitlements may have a decisive effect on the state of the local ecosystem, regardless of what the others wish or do. Therefore, the task of identifying whether and to what extent each is entitled to payment is important but complex.[97] This is all the more so because the existing legal relationships between those with different rights over the land have been developed to deal with established commercial uses and may not be well-equipped to accommodate what is required to ensure the delivery of

---

95 For an analysis of how more positive conservation obligations lead to a new characterisation of property rights, see Christopher P. Rodgers, 'Nature's place? Property rights, property rules and environmental stewardship' (2009) 68 *Cambridge Law Journal* 550.

96 See Chapter 5. All this, of course, assumes that the nature of the rights and who holds them are clear, which may well not be the case, especially in some developing countries.

97 Nsoh and Reid, 'The privatisation of biodiversity' (n. 94) 12.

any newly acknowledged ecosystem services.[98] To the extent that a service relies on restricting the exercise of existing rights, a clear basis for payment can be identified, but delivery of an ecosystem service into the future may also depend on accepting additional positive obligations to look after land in a particular way, and on the overall effect of exercising (or choosing not to exercise) a range of rights over the land, rights that may be held by different parties. These issues are particularly severe in areas characterised by communal ownership and occupation, where the potentially conflicting interests of various parties make it difficult to allocate obligations and entitlements.[99] A further complication is the fact that ecosystem services will often be provided by areas of land that do not constitute a single occupancy unit, for example, depending on all or part of a water catchment, not individual farms or estates. This raises immediate difficulties over identifying who is the provider of the service and what share of benefits or burdens is to be borne by a particular plot.

Similar issues arise under biodiversity offsetting schemes which may allow biodiversity credits to be generated and sold by landowners who commit to enhancing and protecting biodiversity values on their land.[100] Allocating entitlements, liabilities and responsibilities calls for a system that not only makes it easier to share responsibilities and benefits, particularly among those who already hold ownership or occupational rights in areas with multiple occupancies,[101] but also assures buyers and sellers of the security of the rights over the ecosystems services or biodiversity credits they are trading. The new legal rights must take some recognised form, either as a kind of property or as contractual rights,

---

[98] E.g. although in an agricultural context it may make sense to ensure that the occupier always has the right to exercise pest control, in relation to wider wildlife management there may be complexity and fragmentation arising from the fact that in England the Ground Game Act 1880 provides that the tenant as occupier will have the concurrent right to kill rabbits and hares, even though rights to game can be retained by the landowner or passed to someone other than the tenant when land is leased.

[99] E.g. common land, on which see Christopher P. Rodgers and others, *Contested Common Land: Environmental Governance Past and Present* (Earthscan, 2011).

[100] DEFRA, *Biodiversity Offsetting in England* (n. 66); Business and Biodiversity Offsets Programme (BBOP), *Biodiversity Offset Design Handbook* (2012); eftec, IEEP and others, *The Use of Market-based Instruments for Biodiversity Protection: The Case of Habitat Banking, Technical Report for European Commission DG Environment* (eftec, 2010) 29–53.

[101] See e.g., Crofters (Scotland) Act 1993, ss. 50, 50A, in relation to using common grazing land for forestry.

imbued with the characteristics of these different forms of right. The long-term nature of the obligations involved suggest that utilising property-based legal structures is appropriate and important for the success of offset schemes,[102] but these often come with restrictions which limit their flexibility.[103]

There are parallels with the development of carbon credits. Rights to carbon credits that allow carbon emissions and savings to be traded are viewed in some places as separate from the broader rights to land, forests and other immovable property where the relevant operations take place, while in others they remain attached to ownership and occupational rights. For example, the district court in the State of Louisiana in the United States has held that the 'right to report, transfer, or sell carbon credits is part of the bundle of rights associated with property ownership'.[104] This suggests that credits are a feature of holding real property, arising automatically.[105] Other courts, though, have been reluctant to recognise such new rights in credits, concerned that if a carbon credit were recognised as property, then the state would have to compensate the owner if it ever revoked any carbon credits.[106] Similarly in the United

---

[102] The need for definable property rights is a key requirement for offsets identified in Anthea Coggan, Edwin Buitelaar, Jeff Bennett and Stuart M. Whitten, 'Transferable mitigation of environmental impacts of development: two cases of offsets in Australia' (2013) 15 *Journal of Environmental Policy Planning* 303.

[103] Such as the common law's rejection of enduring obligations other than for the benefit of neighbours as noted above (text at n. 96) or the *numerus clausus* doctrine in civil law.

[104] *Roseland Plantation, LLC* v. *US Fish and Wildlife Service et al.*, 2006 US Dist LEXIS 29334 (WD La. 2006) [10].

[105] *Ibid.* [8]. Similarly, in an unpublished opinion in the State of California, the court found that a claimant (Kaiser) could claim ownership of emission reduction credits generated from its use of leased equipment, because the possessory interest in leased equipment entitled Kaiser to operate the equipment, which operation produced emissions: *Kaiser International Corp.* v. *Hearing Board of the South Coast Air Quality Management District et al.* (2006 Cal App Unpub LEXIS 3135) [21].

[106] John Monterubio, 'Recognition of property rights in carbon credits under California's New Greenhouse Gas Cap-and-Trade Program' (2012) 12 *Sustainable Development Law and Policy* 32, 64–5; Markus Gehring and Charlotte Streck, 'Emissions trading: lessons from SOx and NOx emissions allowance and credit systems legal nature, title, transfer, and taxation of emission allowances and credits' (2005) 35 *European Law Reporter* 10,219; Travis Allan and Kathy Baylis, 'Who owns carbon? Property rights issues in a market for greenhouse gases' (2006) 7 *Current Agriculture, Food and Resource Issues* 104.

Kingdom, fishing quota allocations have been recognised as a form of property,[107] but do not enjoy the same level of protection and priority as many other property rights.[108] Pragmatic concerns over the consequences in the wider context (especially any constitutional protection for property rights), rather than the application of the standard tests for identifying when entitlements qualify as property rights, seem to be the driving force behind the denial of full property status to credits.

These decisions reveal the tension and potential for mismatches as 'rights' are perceived from different viewpoints. Private entitlements are shaped by the regulatory context[109] and some form of 'rights' is necessary for transactions to occur. From the private law perspective it seems odd not to recognise these as property. Yet that label carries further consequences in terms of how they are viewed constitutionally, consequences which may be seen as having such undesirable implications that they should be avoided. Stopping short of recognising credits as property may, however, hinder the development of a fully effective network of rights which will enable trading to operate effectively and raise doubts over how credits and entitlements can be treated. What is the status of such 'not-quite-property' rights in relation to insolvency, use as securities, and transferability in terms of formalities and guarantees of title?

Recent litigation in relation to the European Union Emissions Trading Scheme (EU ETS) shows that a failure to define the legal status of the credits can have a major impact on regulatory certainty, business confidence and environmental protection.[110] For example, in *Armstrong DLW GmbH* v. *Winnington Networks Ltd*,[111] which was a case involving fraud and 'stolen' carbon credits, one of the issues that led to litigation was the lack of certainty as to the rights and obligations of the holders of carbon credits, which partly stemmed from the failure to appreciate the legal nature of such intangible property within the EU ETS. Applying Lord Wilberforce's definition of a property right in *National Provincial Bank*

---

[107] As a 'possession' in terms of the European Convention on Human Rights.

[108] *United Kingdom Association of Fish Producer Organisations* v. *Secretary of State for Environment, Food and Rural Affairs* [2013] EWHC 1959 (Admin) [113].

[109] Eloise Scotford and Rachael Walsh, 'The symbiosis of property and English environmental law: property rights in a public law context' (2013) 76 *Modern Law Review* 1010.

[110] *Armstrong DLW GmbH* v. *Winnington Networks Ltd* [2012] EWHC 10 (Ch), [2013] Ch. 156. The EU ETS is the first and the largest regional greenhouse gas emissions trading scheme in the world.

[111] *Ibid.*

Ltd v. *Ainsworth*,[112] this case found that European Union Allowances (EUAs) (carbon credits) were a form of (intangible) property under common law.[113] Until then, as Low and Lin have argued, 'one of the critical failures of the EU ETS lies in its failure to properly define the fundamental legal nature of the EUA'.[114] Clarity as to the legal nature of such credits and entitlements therefore needs to be properly addressed in designing new structures for dealing with biodiversity. Characterisation as property rights may provide a useful shorthand or default position for dealing with some of the issues of certainty, but should not be an end in itself. As Lucas has noted in a different context:

> While there is some value ... in analysing the legal character of rights to carbon sequestration potential and sequestered carbon, the core questions should not be legal characterization of the right but rather how the property right should be structured in functional terms to most effectively support a marketable carbon offset right.[115]

The same applies to designing new structures for dealing with biodiversity.

A different way of using property rights is the revival of the public trust doctrine that is being seen in the United States.[116] This approach

---

[112] [1965] AC 1175, 1247–8.

[113] *Armstrong DLW GmbH* (n. 110) [50]–[53]. Such status is important, e.g. in relation to protection under insolvency rules: *Celtic Extraction Ltd (In Liquidation), Re Bluestone Chemicals Ltd* v. *Environment Agency* [2001] Ch. 475; [2000] 2 WLR 991.

[114] Kelvin F.K. Low and Jolene Lin, 'Carbon credits as EU like it: property, immunity, tragiCO$_2$medy?' (2015) 27 *Journal of Environmental Law* 377, 381–6.

[115] Alastair Lucas, 'The significance of property rights in biotic sequestration of carbon' in Aileen McHarg and others, *Property and the Law in Energy and Natural Resources* (OUP, 2010) 452. See also in the same volume Lee Godden, 'Governing common resources: environmental markets and property in water' (413).

[116] Joseph Sax, 'The public trust doctrine in natural resource law: effective judicial intervention' (1970) 68 *Michigan Law Review* 471; Raphael Sagarin and Mary Turnipseed, 'The public trust doctrine: where ecology meets natural resources management' (2012) 37 *Annual Review of Environment and Resources* 473; Michael C. Blumm and Aurora Paulsen, 'The public trust in wildlife' (2013) 6 *Utah Law Review* 1437; Reed Watson, 'Public wildlife on private land: unifying the split estate to enhance trust resources' (2012–13) 23 *Duke Environmental Law and Policy Forum* 291.

takes seriously the assertion that wildlife is owned by the state[117] and, by accepting that wildlife has a place within rather than outside the legal frameworks, it allows rights and obligations to be recognised, able to compete against those of other property owners. The exact extent and consequences of wildlife being subject to a public trust may be open to argument in specific circumstances,[118] but the overall effect has been summarised as requiring the state to act as follows:

> Based on existing public trust doctrine case law, the state must: (1) consider the potential adverse impacts of any proposed activity over which it has administrative authority; (2) allow only those activities that do not substantially impair the state's wildlife resources; (3) continually monitor the impacts of an approved activity on the wildlife to ensure preservation of the corpus of the trust; and (4) bring suit to enjoin harmful activities and/or to recover for damages to its wildlife under the *parens patriae* doctrine.[119]

The effectiveness of this approach in securing any protection depends heavily on the willingness of the state to act in its capacity as trustee, but the basis of its role is transformed, relying on its entitlements as trustee or property holder, rooted in private law, not its regulatory power in public law. Property concepts may thus become significant in protecting nature from harm rather than leaving it exposed.

2.10 SECURITY, MONITORING AND ENFORCEMENT

Establishing the recognition and legal status of the rights involved is one aspect of a wider issue of giving the parties involved in any transactions the security they need that the credits or entitlements they are dealing with will be respected. This issue, like so many here, is complicated by the long time-scales involved; ensuring respect for the rights requires

---

[117] O. Houck, 'Why do we protect endangered species, and what does that say about whether restrictions on private property to protect them constitute takings' (1994–95) 80 *Iowa Law Review* 297.

[118] D. Musiker, T. France and L. Hallenbeck, 'The public trust and *parens patriae* doctrines: protecting wildlife in uncertain political times' (1995) 16 *Public Land Law Review* 87; P. Redmond, 'The public trust in wildlife: two steps forward, two steps back' (2009) 49 *Natural Resources Journal* 249; Mary Wood, *Nature's Trust: Environmental Law for a New Ecological Age* (CUP, 2013); Edward Fitzgerald, 'The Alaskan wolf war: the public trust doctrine missing in action' (2009) 15 *Animal Law* 193.

[119] Musiker, France and Hallenbeck, 'The public trust and *parens patriae* doctrines' (n. 118) 115.

enduring mechanisms that can be relied on to work decades into the future, when the original parties may no longer be around. Inherent in the idea of using private arrangements as the basis of conservation action (such as conservation covenants or contractual payments for ecosystem services) is that the parties can be left to themselves to monitor and enforce their rights as they think fit, rather than this being a matter of state intervention. There is, however, a legitimate public interest in ensuring that where public goals are being served through such mechanisms they are, and continue to be, properly observed and enforced.[120] This is especially the case if public money is involved (through tax breaks, subsidies or direct payments) and even more where conservation measures achieved through these mechanisms are a condition of obtaining an official permit.[121] Left wholly to their own devices, the parties over the years may lose the desire or capacity to keep a close eye on each other and there is a danger of agreements becoming dormant. Equally, unless proper procedures are put in place, there is a risk of parties finding themselves without the means to take effective action to protect their rights. There are several aspects to be considered.

In the first place, parties contemplating entering any arrangement such as buying or selling a biodiversity offset credit will want to be assured that what they are dealing with is genuine.[122] This can be based on trust between individual actors or can be supported in some way. Dealing through a broker (discussed below) might be one way of seeking reassurance that a transaction has been properly constructed and will deliver what is being promised, so long as one can be confident that the brokers themselves are reliable and checking the good faith and standards of the parties and projects they are handling. This might call for an accreditation scheme of some sort, established officially or by the industry, to identify bodies that can be trusted. Accreditation could also figure at an individual level, for example, requiring that the valuation of a biodiversity credit or ecosystem service is carried out by an accredited assessor. As an example, under the New South Wales BioBanking

[120] Robert L. Glicksman, 'Regulatory safeguards for accountable ecosystem service markets in wetlands development' (2014) 62 *Kansas Law Review* 943, 956–7; Robert L. Glicksman and Thoko Kaime, 'A comparative analysis of accountability mechanisms for ecosystem services markets in the United States and the European Union' (2013) 2 *Transnational Environmental Law* 259.

[121] E.g. as a condition of obtaining development consent or under the US Clean Water Act 1977 (see Chapter 4, section 4.4).

[122] At the very least any system should seek to reduce the risk of fraud such as that which led to litigation in *Armstrong DLW GmbH* (n. 110).

scheme assessors are required to complete a four-day training course on the BioBanking Credit Calculator before they can start work.[123] An alternative of requiring individual transactions to be approved by an official body would be possible but cumbersome.[124]

Long-term security is helped if there is a clear record of the commitments, obligations and entitlements which have been created. One solution is to introduce some form of official registration, but the form and indeed existence of this will depend largely on the background in the jurisdiction concerned. Where registration of title is already an established part of land law, or where registration is a requirement for agreements to take effect so as to bind successors in title, then the recording of offset agreements, conservation covenants, etc. will seem obvious.[125] On the other hand, where this is not the norm, there may be some objections, as there are in some quarters in the United States to registration of conservation easements, in part because of fears that public access to the information that land has been committed to conservation in this way might attract unwanted attention and visitors of various sorts.[126] The issue of registration is also relevant to coherence, as a means of being able to obtain an overview of the many individual arrangements being made in order to assess their overall effect.[127]

The next aspect to consider is monitoring compliance with whatever terms have been agreed. Although under private law arrangements this is essentially a matter for the parties, it is important that they are in a position to carry out this task. Accordingly, consideration should be given to the right to have access to relevant information and rights of entry to carry out inspections. These can either be left as a matter to be negotiated

---

[123] See New South Wales Government, *Credit Calculator for Major Projects and BioBanking* (5 November 2014), available at www.environment.nsw.gov.au/biobanking/calculator.htm.

[124] Cf. the detailed and complex arrangements for the approval of projects within the Clean Development Mechanism under the Kyoto Protocol set out in the *CDM Rulebook*, available at http://cdmrulebook.org.

[125] See e.g., the discussion of registration in the context of proposals to allow the creation of conservation covenants in England: Law Commission, *Conservation Covenant: A Discussion Paper*, No. 211 (2013) 59–61 and *Conservation Covenants*, No. 349 (2014) 85–91.

[126] James Olmsted, 'The invisible forest: databases and the end of the clandestine conservation of natural lands' (2011) 74 *Law and Contemporary Problems* 51.

[127] See discussion of a central recording scheme in relation to proposed conservation covenants in England; Law Commission, *Conservation Covenant: A Discussion Paper* (n. 125) 60–61 and *Conservation Covenants* (n. 125) 88–91.

individually (perhaps supported by guidance), or be provided for by law in the form of more general mandatory or default provisions. Where an agreement is connected with an official permitting scheme, then the relevant official body may be given a power or duty to monitor compliance.[128] A further issue is whether monitoring should be required as a means of ensuring that agreements are kept alive, with some means of checking up on this. For example, the law on conservation easements in Maine was amended in 2007 to introduce, among other reforms, a legal obligation on easement holders to monitor and report on the state of the land affected every three years.[129] Such provisions seek to ensure that an easement is not forgotten over time. Similarly, providing monitoring reports may be required where offsetting is a condition of obtaining a permit of some sort.[130] Of course, having structures for monitoring is only part of the story and these must be put into practice; a recent survey of staff involved in the town and country planning system in England revealed low levels of compliance monitoring, despite wide statutory powers to assist this.[131]

Ensuring monitoring that identifies non-compliance is one thing, but securing effective enforcement is another. Standard remedies under the law of contract or property tend to favour compensation in financial terms when rights are breached, but that approach misses the point where the whole purpose of an arrangement is to secure continuing biodiversity benefits. The usual alternative is an injunction or interdict, ordering parties in actual or threatened breach of their obligations to comply with these, but in common law jurisdictions (although probably less so in civil law ones) there has been great reluctance on the part of the courts to grant injunctions which will require continuous supervision by the courts into the future.[132] Accordingly, whereas injunctions might be obtained to prevent specific harmful acts there may be doubts over their availability to guarantee the continuing, and even continuous, management steps

---

[128] E.g. under the US wetland mitigation scheme, the district engineer has the power to conduct site inspections to monitor offset sites; 33 CFR §332.6(a)(2), 40 CFR §290.96(a)(2).

[129] Jeff Pidot, 'Conservation easement reform: as Maine goes should the nation follow?' (2011) 74 *Law and Contemporary Problems* 1, 11–16.

[130] E.g. under the US wetland mitigation scheme; 33 CFR §332.6(c), 40 CFR §290.96(c).

[131] A. Sheppard, S. Britnell, J. Cooke and Royal Town Planning Institute, *Planning Enforcement England: At the Crossroads, Project Report* (University of the West of England, 2014), available at http://eprints.uwe.ac.uk/24531/.

[132] *Co-Operative Insurance Society Ltd* v. *Argyll Stores (Holdings) Ltd* [1998] AC 1, 11–16 (Lord Hoffman).

which are needed to maintain the biodiversity or ecosystem services on a particular site, potentially leaving a gap in enforcement. In the context of management agreements with statutory bodies, one solution has been to provide for direct action by the body (with recovery of costs), or in extreme cases compulsory acquisition of the land from the defaulting party,[133] but this may be seen as too radical a step to take in relation to arrangements between private parties.

A further issue arises if the parties for whatever reason do not in fact take steps to enforce an agreement even though its terms are being broken. Given the public interest in the biodiversity goals, a case can be made for there being some means of holding the parties to account. One option which is used in Scotland in relation to conservation burdens (the same as conservation covenants) is to limit those who can hold the benefit of such obligations to a limited number of parties who by their nature as public or charitable bodies can be held to account, legally or politically, by their members or the public.[134] In this way the continuing engagement of the body is taken on trust, but within a wider framework where there is a general mechanism for the accountability of the body, rather than there being detailed provisions seeking to ensure that specific conservation obligations are not overlooked.[135] A second option is to give some official body the standing and powers to intervene. The reform of conservation easements in Maine again provides an example here, granting a power for the Attorney-General to intervene where the holder of a conservation easement has not taken action to respond to a breach of an easement's terms.[136] A more radical solution would be to enable any third party[137] (or perhaps any third party who can show that they have some interest in the matter as oppose to being a 'mere busybody')[138] to

[133] E.g. in relation to Sites of Special Scientific Interest in England and Wales; Wildlife and Countryside Act 1981, ss. 28K, 28N.

[134] See Chapter 4, section 4.4.

[135] Similarly, those entering bio-banking agreements in New South Wales must satisfy the Minister that they are 'fit and proper persons'; Threatened Species Conservation Act 1995, s. 127D; Threatened Species Conservation (Biodiversity Banking) Regulations 2008, Reg. 291 of 2008, reg. 15.

[136] Pidot, 'Conservation easement reform' (n. 129) 11–16.

[137] This broad approach is taken in some Australian contexts such as in New South Wales, where any person can take court action to enforce a 'biobanking agreement', 'whether or not any right of the person has been or may be infringed by or as a consequence of the breach'; Threatened Species Conservation Act 1995, s. 127L.

[138] The rules on standing vary greatly between jurisdictions, but with a general trend towards a more generous approach on environmental matters,

intervene and either enforce the obligation directly or seek to require a party to the arrangement to do so. This, though, might be seen as inconsistent with the private, voluntary nature of the sort of arrangement being discussed and therefore incompatible with the approach here.[139]

2.11 INFORMATION

Before looking at some issues affecting the overall workings of any schemes, three further aspects of the detailed operation of a 'privatised' system need to be considered: information, transaction costs and the nature of payments. Knowing what biodiversity benefits can or will be provided is essential for judging the acceptability of a conservation obligation of any sort and in the assessment of any payments to be made. Yet our knowledge of biodiversity is very limited. Even in a country like the United Kingdom, where there is a long history of gathering 'natural history' records, much of it the product of work by volunteers,[140] our knowledge even of the basic distribution and populations of species is far from complete, far less our understanding of the countless interactions from microbial level upwards which are essential to the health and sustainability of an ecosystem. For the smaller and less attractive species, distribution maps often reveal a distribution of recording effort rather than true distribution,[141] and in less heavily populated areas knowledge may be sparse. The results of site surveys are also subject to variation and uncertainty, depending on methodology, expertise and how the variability within the natural world is treated, all of which can determine

---

although still alert to the risk of abuse by busybodies; see e.g., the comments made in the context of Scotland's previously strict view of standing by Lord Hope in *Walton* v. *Scottish Ministers* [2012] UKSC 44, 2013 SC (UKSC) 67, [151]–[154].

139 Law Commission, *Conservation Covenants* (n. 125) 69–72.

140 The pioneering work of Gilbert White's observations in *The Natural History and Antiquities of Selborne* (1789) continues in individual and collective projects. One example is Dawn Balmer and others, *Bird Atlas 2007–11* (British Trust for Ornithology, 2013), an atlas of the distribution of bird populations in the British Isles, drawing on data from over 40,000 people, with over 17,000 directly participating in the collecting and entering of data (*ibid.* 10).

141 Even for common and widespread but often overlooked species it can be possible to add to the confirmed distribution simply by reporting findings to the relevant recording scheme, as one of the authors of this book has done with common species of woodlice in several parts of Scotland.

whether certain species are or are not recorded as being present.[142] The fundamental basis for taking decisions on valuation may therefore be lacking. Such uncertainties in knowledge are exacerbated when there is a lack of expertise among the parties to understand, analyse and assess the data that is available. In the United Kingdom, for example, there is concern that too few local authorities have any ecologically trained staff, even though understanding the biodiversity issues is vital for carrying out their legal responsibilities for assessing biodiversity issues in the context of decisions on development projects and under proposals for biodiversity offsetting.[143]

Where information is held, it may not be evenly shared. Asymmetry of information, where one party to a transaction has more knowledge than the other, is a widely recognised form of market failure but is very likely in any biodiversity transactions.[144] The landowner is the party with the background of involvement with the site and who has control over access so as to enable or restrict detailed investigations. For large-scale features the landowner may have better knowledge, but the tables can be reversed for less obvious features where only a few scientists might be able to identify the presence of ecologically important but otherwise insignificant plants or invertebrates. Even where data are in the public domain, parties may lack the ecological understanding to appreciate its significance and the factors that are important for securing the continuing health or enhancement of the ecosystem.

Transactions may therefore be contemplated on the basis of a far from complete or equally shared understanding of the true position. In such circumstances there may be greater reliance on accreditation schemes such as discussed above to give the parties confidence in the factual basis on which business is being done. A further option is to design schemes which allow the varying knowledge to play a part, such as auctioning schemes whereby landowners bid for contracts, informed by their own

---

142 See e.g., the experiences related in Morgan Robertson, 'The nature that capital can see: science, state and market in the commodification of ecosystem services' (2006) 24 *Environment and Planning Development: Society and Science* 367.

143 UK All Party Parliamentary Group on Biodiversity, *Biodiversity APPG September 2014 Update*, Appendix 1.

144 OECD, *Harnessing Markets for Biodiversity: Towards Conservation and Sustainable Use* (OECD, 2003) ch. VII; Jason Shogren, Rodney Smith and John Tschirhart 'The role of private information in designing conservation incentives for property owners' in Jason Shogren (ed.), *Species at Risk: Using Economic Incentives to Shelter Endangered Species on Private Lands* (University of Texas Press, 2005).

assessment of the state of their land. For example, the auction-based approach used in the BushTender programme in the State of Victoria, Australia, and in the Fowey River Improvement Auction pilot project in England, relies on a publicised competition among landholders who provide sealed bids to the government or private entity of how much they are willing to accept for specified land use management changes informed by their own assessment of the state of their land and what is needed to deliver specified benefits.[145] The process therefore places far greater emphasis on landowners to initiate participation in the schemes.

2.12 TRANSACTION COSTS AND INTERMEDIARIES

Acquiring and sharing information is one element that must be borne in mind when the costs of implementing schemes based on trading or greater private initiative are considered. For a market to work smoothly transaction costs must be kept low enough to make trading worthwhile.[146] Therefore, in creating new markets such costs must be kept low enough to stimulate an active market and thus gain the benefits that can flow from this approach. Costs arise and affect various stages in the process,[147] and vary according to the nature of 'goods' being traded, the actors involved and the institutional arrangements.[148] Identifying other

---

[145] See Gary Stoneham and others, 'Auctions for conservation contracts: an empirical examination of Victoria's BushTender trial' (2003) 47 *Australian Journal of Agricultural Resoures Economics* 477; CSERGE and Westcountry Rivers Trust, *Payment for Ecosystem Services Pilot Project: The Fowey River Improvement Auction, Final Report Submitted to the Department for Environment, Food and Rural Affairs* (19 September 2013). See Chapter 3, section 3.5.3.

[146] In the United States, transaction costs have been identified as a factor affecting the use of conservation easements (Dominic P. Parker, 'Land trusts and the choice to conserve land with full ownership or conservation easements' (2004) 44 *Natural Resources Journal* 483), whilst fears of higher transaction costs when using the bio-banking scheme rather than alternative pathways have been reported as a disincentive to participation in this scheme within New South Wales, as has the time taken to review and process bio-banking proposals (New South Wales Government, *Biobanking Scheme: Statutory Review Report* (2014) 9–11, 14).

[147] Anthea Coggan, Stuart Whitten and Jeff Bennett, 'Influences of transaction costs in environmental policy' (2010) 69 *Ecological Economics* 1777; Arild Vatn, 'An institutional analysis of payments for environmental services' (2010) 69 *Ecological Economics* 1245, 1248–9.

[148] Ronald Coase, 'The problem of social costs' (1960) 3 *Journal of Law and Economics* 15.

parties in a position to trade is an initial task, so that developers seeking to offset the biodiversity impact of a project must be able to contact those able to provide an offsetting project that can fill their needs. There must be information readily available and an accepted valuation system to enable both parties to be satisfied that a fair exchange is being achieved. Any formalities to provide security and certainty in the deal must be proportionate, whilst the costs of any market oversight must also be considered. The more complex, detailed and rigorous any processes are, the longer and more expensive they are likely to be, increasing costs both in financial terms and in terms of the effort, time and resources required before progress can be made. A choice may have to be made between a simple, clear and quick system and one which is more sophisticated and flexible but for those very reasons less inviting to participants.[149] The solutions to many of the challenges noted above will thus have an influence on the cost of transactions.[150]

Transaction costs can also be affected in other ways. The size of the market is one aspect which has an influence.[151] If a scheme is voluntary, entering the market and bearing the costs involved are optional, and with the likelihood of fewer participants, finding a suitable counterpart may be more difficult. On the other hand, a mandatory scheme creates a busier marketplace where lots of parties are available and need to do business, providing the basis and incentives for greater efficiency at all stages in transactions, and building up the experience to be able to deliver it.[152]

A further option is to make use of an intermediary who can help the parties wanting to trade to find partners and navigate the various stages of

---

149 See e.g., the debates over appropriate ways of valuing credits for offsetting schemes; see Chapter 4, section 4.3.

150 Anthea Coggan and others, 'Factors that influence transaction costs in development offsets: who bears what and why?' (2013) 88 *Ecological Economics* 222; Coggan, Whitten and Bennett, 'Influences of transaction costs in environmental policy' (n. 147).

151 In New South Wales, it was noted that the site assessment costs were a disincentive to participating in the bio-banking scheme but that '[a]s participation rates in the scheme have increased, competition among assessors has begun to drive survey costs down'. New South Wales Government, *Biobanking Scheme: Statutory Review Report* (2014) 14, available at www.environment.nsw.gov.au/biobanking/bbfindings.htm.

152 See discussion in the context of the Biodiversity Offsetting pilot in England; House of Commons Environmental Audit Committee, *Biodiversity Offsetting*, HC 750 (2013–14) 14–16.

a deal. Various types of intermediary can be identified.[153] Information providers, brokers (who will try to match parties as well as providing information) and aggregators (who will take on certain responsibilities for parties on the basis of their knowledge of and access to a wide range of projects) can become involved on an unofficial basis, whereas a clearing house may need more formal authorisation, whilst a bio-bank may take direct ownership and responsibility for providing an offset or service.[154] All of these possibilities create different networks of responsibilities and potential liabilities between parties, generating a need to think carefully about who bears the costs and risks involved, questions complicated by the long time-scales and uncertainties affecting so many biodiversity projects. Although fees of some sort may be involved, overall the use of intermediaries may well help to reduce transaction costs for parties as opposed to leaving them to find and negotiate directly with other parties in a market which is novel and relies on a degree of specialist knowledge.[155]

2.13 INPUT OR OUTPUT PAYMENTS

Where habitat (re-)creation or enhancement is involved, the time-lag noted above between the initial work (when most costs are likely to be incurred) and the eventual delivery of biodiversity gains may be substantial.[156] The question therefore arises of when any reward should be paid to the landowners concerned. It is asking a lot of landowners to incur a financial loss today (whether direct expenditure or lost income) to begin the process and then wait for years before anything is given in exchange, at the point when the biodiversity benefits start to materialise. This is even more the case since the uncertainty of all natural systems means that no-one can be certain of exactly what will be achieved. Yet to give the reward 'up front' risks the money being wasted if the project is abandoned before it produces real gains or if it is unsuccessful. Such payment for the initial work done, rather than for the results, also gives no reward to those responsible for the continuing management of the land at the time when the gain or service is actually being delivered. Early

---

[153] Anthea Coggan and others, 'Intermediaries in environmental offset markets: actions and incentives' (2013) 32 *Land Use Policy* 145.

[154] See Chapter 4, sections 4.4. and 4.7.

[155] Coggan and others, 'Intermediaries in environmental offset markets' (n. 153).

[156] See also Chapter 3, sections 3.2 and 3.3.

payment means that so-called payment for ecosystem services schemes are not really paying for the delivery of services, but for work that it is hoped will deliver services in the future. The basis for calculating payments is also affected if the focus is on the initial costs of, for example, planting a hedgerow, rather than the level of services actually being delivered, such as the number and density of species using the hedgerow.

Even though it may undermine the link between payment and delivery of an ecosystem service, the preference for 'input' rather than 'output' payments has a pragmatic basis. It is simpler for all the parties to deal with payments over a shorter time-span. Public bodies and other organisations cannot usually commit to or budget for funding obligations years into the future. The immediate reward comes at a time when there are expenses being incurred (which would otherwise have to come from the landowners' own resources or lending) and strengthens the incentive to join a scheme since the reward is immediate and guaranteed and the owners are never 'out of pocket'. The risk of things not turning out as expected is borne by the funding body, not the landowners, again making involvement less risky and more attractive for the latter. For these reasons, input payments are widespread, although there are some examples of schemes that do pay on an output basis, such as the geese management schemes paying compensation based on geese numbers to farmers for losses caused by the grazing of overwintering geese in parts of Scotland.[157]

There are endless variations which can endeavour to balance the advantages of an input approach with ensuring that delivery of gains or services is achieved and supported. Initial payments to support the most costly early stages can be complemented with continuing payments to support the subsequent management activities, or payment can be phased, with the later payments dependent on acceptable progress with the enhancement of the habitat or whatever.[158] There can be claw-back

---

[157] Scottish Natural Heritage, *Islay Sustainable Goose Management Strategy October 2014–April 2024* (2014), available at www.snh.gov.uk/docs/A1434517.pdf.

[158] E.g. at one time in the United Kingdom the grants to support afforestation projects were paid on the basis of 70 per cent being paid when planting was finished, 20 per cent when adequate stocking was achieved and 10 per cent five years later, with separate schemes coming into effect after that stage to support continuing management; Colin T. Reid, 'Forestry, the law and the environment' in Christopher P. Rodgers (ed.), *Nature Conservation and Countryside Law* (University of Wales Press, 1996) 151.

provisions if it is found that landowners are not in fact observing the conditions under which payments have been made.[159] The trend, though, has been towards simpler, shorter-term schemes, reducing the future commitment of funds and the commitment to continuing monitoring. As in so many instances, though, the key point is to be aware of the issue and to bear it in mind as design choices are being made in developing a scheme, or group of related schemes, to support action taken to further biodiversity.

2.14 GOVERNANCE

Whatever approach is taken, issues of governance will arise. A shift towards market-based mechanisms changes rather than removes the role for government, moving the focus from regulating activities with direct impacts on biodiversity to regulating the market which shapes how others affect biodiversity. The need for regulation does not go away, as shown by experience in the more established field of carbon markets: 'The main lesson for the regulation of offset markets is that the markets should be more tightly controlled than they were initially'.[160] Governance is at times a rather slippery term and covers a wide area,[161] but key components of good governance include transparency, accountability and public participation[162] and these must be considered in the design and operation of any conservation scheme. What is seen as most important, and how issues will be resolved, will vary greatly,[163] depending on the broader governmental and legal context, but there must be a means of

---

[159] In 2008 one Scottish landowner had to repay £107,650 of the agricultural subsidies paid to him following breaches of the environmental conditions for such financial support, in this case the discovery of illegal pesticides connected with the poisoning of birds of prey; (2008) 405 *ENDS Report* 53.

[160] Christopher Arup and Hao Zhang, 'Lessons from regulating carbon markets' (2015) 4 *Transnational Environmental Law* 69, 99.

[161] Elizabeth Fisher, Bettina Lange and Eloise Scotford, *Environmental Law: Text, Cases and Materials* (OUP, 2013) ch. 13.

[162] E.g. the Principles for the Advancement of Justice, Governance and Law for Environmental Sustainability, contained in the Rio+20 Declaration on Justice, Governance and Law for Environmental Sustainability agreed at the World Congress on Justice, Governance and Law for Environmental Sustainability, Rio de Janeiro, 17–20 June 2012, available at www.unep.org/rio20/Portals/24180/Rio20_Declaration_on_Justice_Gov_n_Law_4_Env_Sustainability.pdf.

[163] James Salzman and J.B. Ruhl, 'Currencies and commodification of environmental law' (2000–01) 53 *Stanford Law Review* 607, 668–93.

ensuring that schemes are widely accepted as having the legitimacy to enable them to continue, and especially to justify any advantages granted to participants, whether financial or in terms of legal privileges.[164]

A conservation policy based more strongly on privatised techniques might lack transparency. If conservation is in the hands of a statutory body, then it is likely that through annual reports and other publications it will be possible to find out what is being done. If more conservation activity transfers to the private sector through easements, offsets or ecosystem service payments, not only will there not be a single focus for gathering and providing information, but much of the data may be held outwith the scope of statutory reporting rules, and of any obligations on public authorities to provide access to environmental information.[165] This may call for further measures to be put in place to ensure the collection of data on,[166] and publicity for, the steps being taken.[167]

Accountability too is affected if more responsibility passes from the work of centralised bodies to the cumulative impact of lots of individual transactions. Under such a more fragmented system, it is harder for anyone to be held to account if the system fails to achieve its objectives. To overcome this, some oversight body might be put in place, but establishing any structure that gives such a body a grip strong enough to be accountable must at the same time diminish the flexibility and distributed decision-making that is a strength of moving to a more privatised approach. Ensuring accountability may mean that public authorities have to retain a strong role, albeit increasingly as regulator of others rather than as direct provider of conservation activities.

Linked to both of the above is the issue of public involvement, where there are two dimensions, direct involvement and wider public participation. Each of these in turn provides means of giving effect to the ideas of shared responsibility[168] and stewardship (especially in relation to

---

[164] Such as departures from the standard rules against enduring restrictions on land; see the discussion of covenants in Chapter 5.

[165] As under Directive 2003/4/EC of the European Parliament and of the Council of 28 January 2003 on public access to environmental information and repealing Council Directive 90/313/EEC, [2003] OJ L41/26.

[166] There is obviously a link here with any registration steps required in relation to the creation and even validity of any obligations which are tied to specific land; see Chapter 4, section 4.6 and Chapter 5, section 5.6.

[167] A study in Australia has identified greater transparency as a key element in avoiding negative effects of offsetting schemes; Gordon and others, 'Perverse incentives risk undermining biodiversity offset policies' (n. 42).

[168] Declaration of the United Nations Conference on the Human Environment, Stockholm 1972, UN Doc. A/Conf.48/14/Rev 1 (1973), Preamble, para. 7

land)[169] which are features of a move to a more sustainable future. The new approaches under consideration here offer more by way of direct involvement to a wide range of participants than does the more traditional approach relying on public bodies dominating conservation activity. At present it is the authorities who determine which species and habitats are to be protected and the form that protection is to take, and who provide the resources to support most of the conservation activity. Conservation covenants and biodiversity offsets open up to those on both the supply and the demand sides of the equation the possibility to select for themselves the particular areas of land and sorts of habitats to be protected and how this is to achieved. Payment for ecosystem services may not offer any choice to those who have to pay, but does at least connect more people directly with the natural cycles that affect and are affected by them. Payment may also offer an incentive to conserve the natural environment for those outwith the hotspots that are the focus of policy dedicated to protecting the most threatened aspects of biodiversity. In these ways, the ideals of shared responsibility and stewardship are brought into play, as opposed to conservation being a matter that is dealt with exclusively by the public authorities. The involvement of more parties also offers opportunities, albeit also challenges, for a greater integration between conservation and other aspects of activity and policy, a further desirable element of good environmental governance.[170] Additionally, it is possible to utilise more directly and effectively the wealth of tangible and moral support for nature conservation that is currently directed through wildlife charities and other bodies which find themselves confined to a legally restricted, albeit often important, role.[171] There are, though, concerns that such developments might increase the risk that voluntary efforts and contributions are diverted towards meeting

---

[169] William Lucy and Catherine Mitchell, 'Replacing private property: the case for stewardship' (1996) 55 *Cambridge Law Journal* 566; for a detailed analysis of different concepts of stewardship, see Emily Barritt, 'Conceptualisng stewardship in environmental law' (2014) 26 *Journal of Environmental Law* 1.

[170] Tim Jewell and Jenny Steele (eds), *Law in Environmental Decision-Making: National European and International Perspectives* (Clarendon Press, 1998) 3–9.

[171] In the United Kingdom, this importance is reflected by the potential for sites managed by a range of bodies to be legally recognised as National Nature Reserves; Wildlife and Countryside Act 1981, s. 35.

legal conservation obligations rather than making an additional contribution beyond the legal requirements.[172]

When it comes to the participation of the public as stakeholders, there is a danger that public participation may favour the obvious and attractive species and habitats over ones that appear less appealing or significant.[173] Leaving things in the hands of the experts, however, detached from the public's concerns, can be equally disadvantageous.[174] Public participation is not always a part of the process leading up to important conservation decisions,[175] although in practice consultations may be held on many occasions. Specific decisions may be covered by rules on environmental impact assessment or procedures in other approval mechanisms where there is a role for the public (for example, in obtaining development consent), but the underlying point remains that conservation decisions are mostly ones for the authorities to make themselves, with public participation being an incidental consequence of the interaction with other decision-making processes rather than an integral part of conservation regulation.

If new mechanisms were adopted on a widespread basis as the means of achieving conservation goals, the role of the wider public as stakeholders might be even more marginalised. Conservation covenants and biodiversity offsets would be a matter of private negotiation between the parties for whose benefit they are creating the covenants, and capable of discharge by their private agreement. Even though potentially dependent on official approval in some form, at which stage some publicity might

---

172 Gordon and others, 'Perverse incentives risk undermining biodiversity offset policies' (n. 42) 534.

173 The absence of public sympathy for some species was shown by the leading article in *The Times* (London), 4 January 1991, and subsequent letters when the adder was first given statutory protection.

174 See e.g., Royal Commission on Environmental Pollution, *Setting Environmental Standards*, 21st Report, Cm 4053 (1998), especially ch. 7.

175 In most of the United Kingdom, there is no formal role for the public in the designation of the most important conservation sites (Natura 2000 sites, Nature Reserves or Sites of Special Scientific Interest), although in Scotland the designation of an SSSI does require specific notification to a wide range of 'interested parties' who are given the opportunity to make representations and there is also a requirement for publicity in the local press; Nature Conservation (Scotland) Act 2004, ss. 3(1), 48(2) and Sch. 1 (cf. the lesser publicity arrangements for SSSIs in England and Wales under the Wildlife and Countryside Act 1981, s. 28).

be involved,[176] the arrangements would rely on the private decisions and negotiations of those involved. The fact that money is changing hands for ecosystem services would bring arrangements away from the regulatory and into the commercial and financial spheres where public participation is generally more limited at the individual level. By moving away from public law mechanisms, the opportunities for standard approaches to public participation are reduced.[177] Such participation is linked to the direct action or regulatory decisions of a public authority and will diminish if conservation effort relies less on such public intervention and more on the aggregation of many individual decisions in the private sector and of individual transactions between private parties themselves or between private parties and the regulatory bodies.

A final governance issue is the need for coherence with other areas of policy. To the extent that any tax breaks or subsidies are involved, coherence with other aspects of the state's fiscal policies should be ensured and perverse incentives eliminated. Financial and other measures must also be designed so as not to fall foul of the rules affecting competition, state subsidies and aspects of the various regimes designed to ensure a level playing-field in inter-state commerce, common markets and international trade.[178] At a wider level, since so many activities have a direct or indirect effect on biodiversity, care should be taken to remove incoherences across different areas of law and policy. Obvious areas to consider, because of their impact on land, are development planning, agriculture, transport and aspects of energy. The worst examples are, we hope, now in the past, such as when the grants offered by one government body to convert uncultivated land to agriculture had to be taken into account when another body offered payments to prevent this occurring.[179] Nevertheless, the risk of different areas of policy pulling against each other remains and must be checked for.

---

176 Under the US wetland mitigation scheme, there must be some public notice of the project involved, including at least some details of the compensation plans; 33 CFR §332.4(b), 40 CFR §230.94(b).

177 On the role and value of public participation see Frans H.J.M. Coenen (ed.), *Public Participation and Better Environmental Decisions: The Promise and Limits of Participatory Processes for the Quality of Environmentally Related Decision-making* (Springer, 2009).

178 As illustrated, e.g. in the famous *Tuna – Dolphin* and *Shrimp – Turtle* cases; *Restrictions on Imports of Tuna*, GATT, (1991) 30 ILM 1598 and (1994) 33 ILM 839; *Import Prohibition of Certain Shrimp and Shrimp Products*, WTO Appellate Body, (1998) WT/DS58/AB/R, (1999) 38 ILM 118.

179 Christopher P. Rodgers, *The Law of Nature Conservation* (OUP, 2013) 116–21.

2.15 OVERVIEW

The issues discussed in the immediately preceding sections reveal tensions which must be faced in contemplating a shift to greater reliance on private mechanisms to secure biodiversity conservation. The health of our ecosystem is something that is of concern to the whole community, justifying public responsibility and accountability. Yet any mechanisms that rely on private property, private transactions and private initiative necessarily entail a more fragmented, less open and overall less accountable system. Short of the total nationalisation of all land and natural resources, at one extreme, or the complete abandonment of all public intervention in environmental matters, at the other, any approach to environmental matters will represent a compromise. That compromise can be drawn at different points, and often will be in different (and inconsistent) places even within the one jurisdiction. Moving away from the established patterns to adopting a different approach will affect that compromise and this is something that should be considered as reforms are considered.

That any change will involve such a shift in balance is not a reason for not innovating. All of the matters discussed in this chapter are ones that need to be considered, not absolute obstacles which should prevent new approaches being adopted once the implications in terms of such issues have been carefully weighed. The present structures themselves embody a host of decisions on responsibility, flexibility, information, comparative value and how we look at the natural world, decisions which are often not fully appreciated. Contemplating different approaches should turn the spotlight as firmly on the choices embedded in the current frameworks as on what the new mechanisms might bring. No system is perfect and the recognition of weaknesses in new approaches should not stand in the way of a proper assessment of whether improvements on the status quo can be achieved. In many cases a lot depends on the precise design of the mechanisms that come to be adopted and in the following chapters it is time to turn to examine in more detail some key mechanisms and their variations and to explore the potential and risks that they create.

3. Payment for ecosystem services

Payment for ecosystem services (PES) is based on the recognition that land left in an undeveloped state is not actually unproductive, as it is often perceived, but is providing a range of services of great practical, economic and spiritual value to society. The aim is to ensure that those who provide these services are paid for doing so, and in this way, it is argued that the vast cost of replacing them will be saved by providing a financial incentive to ensure their continued provision,[1] in the same way that a water company pays farmers for changes to land management practices to improve water quality, thereby reducing drinking water treatment costs. The idea of 'selling nature to save it',[2] of which PES is a part, has emerged as one of the ambitious concepts considered to enhance the conservation of biodiversity at the national, regional and international levels.[3] In particular, the Convention on Biological Diversity (CBD) considers PES as being a key tool to protect biodiversity,[4] and it is

---

[1] Matthias Schröter and others, 'Ecosystem services as a contested concept: a synthesis of critique and counter-arguments' (2014) 7 *Conservation Letters* 514; Joshua Bishop and Chloe Hill (eds), *Global Biodiversity Finance: The Case for International Payments for Ecosystem Services* (Edward Elgar Publishing, 2014).

[2] Kathleen McAfee, 'The contradictory logic of global ecosystem services markets' (2012) 43 *Developmental Change* 105, 106.

[3] Erik Ansink and Jetske A. Bouma, 'Payments for ecosystem services' in Jetske A. Bouma and Pieter J.H. Van Beukering (eds), *Ecosystem Services: From Concept to Practice* (CUP, 2015); Joshua Bishop and others, 'Introduction to international payments for ecosystem services' in Joshua Bishop and Chloe Hill (eds), *Global Biodiversity Finance: The Case for International Payments for Ecosystem Services* (Edward Elgar Publishing, 2014); Lars Hein, Daniel C. Miller and Rudolf de Groot, 'Payments for ecosystem services and the financing of global biodiversity conservation' (2013) 5 *Current Opinion in Environmental Sustainability* 87; OECD, *Paying for Biodiversity: Enhancing the Cost-Effectiveness of Payments for Ecosystem Services* (OECD, 2010).

[4] Among the CBD's strategies for resource mobilisation is the goal, '[t]o promote, where applicable, schemes for payment for ecosystem services, consistent and in harmony with the Convention and other relevant international obligations': Convention on Biological Diversity (adopted 5 June 1992, entered

also promoted by Agenda 21 established at the 1992 United Nations Conference on Environment and Development, in Rio de Janeiro, Brazil.[5] The Global Environment Outlook (GEO) 5 also identifies PES as one of the policy responses/instruments based on best practice adopted successfully in one or more regions around the world that would speed up the achievement of internationally agreed goals on biodiversity conservation.[6] The identification of PES in these various policy documents as a policy option for the preservation of biodiversity based on best practice around the world suggests that the concept already has widespread recognition at various levels. But its development into an 'acceptable' policy option for nature conservation has not been instant. It has been a gradual process and its evolution could be traced back to the ecosystem approach which involves 'a strategy for the integrated management of land, water and living resources that promotes conservation and sustainable use in an equitable way'.[7] Central to this approach is a shift in our way of thinking and policy-making towards conservation that takes into account the entire ecosystem structure and its functioning. This is based on the increased recognition of the role of the 'natural' environment in providing a range of services, either directly or indirectly.

Four broad categories of ecosystem services are recognised: provisioning services such as food, crops, timber, fresh water and plant-derived medicines; regulating services such as the effect on water quality through the filtration of pollutants by wetlands, climate regulation through carbon storage and water cycling, and protection from disasters and disease; cultural services that provide recreation, spiritual and aesthetic values, and education; and supporting services such as soil formation, photosynthesis and nutrient cycling.[8] While the financial value of some ecosystem services such as food, crops, timber and medicines is recognised in the marketplace (although the market value may not always

into force 29 December 1993) 1760 UNTS 79, COP 9 Decision IX/11 (2008), objective 4.1.

[5] UN GAOR, 46th Sess., Agenda Item 21, UN Doc. A/Conf151/26 (1992), Agenda 21, Chapter 10 Integrated Approach to the Planning and Management of Land Resources.

[6] UNEP, *Global Environment Outlook (GEO) 5: Summary for Policy Makers* (UNEP, 2012) 15.

[7] Convention on Biological Diversity (n. 4), COP 5 Decision V/6 (2000).

[8] The Economics of Ecosystems and Biodiversity (TEEB), *The Economics of Ecosystems and Biodiversity: Mainstreaming the Economics of Nature, A Synthesis of the Approach, Conclusions and Recommendations of TEEB* (2010) 7 and Annex 2 and the sources referred to there. See generally the website of the TEEB project, www.teebweb.org. See Chapter 1, Box 1.2.

adequately reflect their dependence on supporting services like pollination and soil formation), others like climate regulation, water cycling and flood regulation are not, despite their vital contribution to human well-being. These regulating and cultural services have historically not been traded in the marketplace but rather publicly funded – or in many cases taken for granted. The invisible nature of the benefits of these ecosystem services has meant that, in the past, decisions affecting the natural environment have often been taken without due consideration given to all these vital services that the environment produces. That said, the value of these services is increasingly being recognised, for example, the health benefits from cultural services, and social and economic benefits from flood regulation. However, with public funding increasingly under pressure, PES aims to ensure that the value of all these ecosystem services is reflected in decision-making,[9] and that those who benefit from these services are in fact paying for them, rather than having the public fund the services.

3.1 DEFINITIONS

While the basic idea of PES may appear straightforward, existing definitions for PES are multifarious and ongoing discussion in the literature identifies two broad categories of definition: Coasean type and Pigouvian type PES.[10] The Coasean type PES is based on the theory developed by the economist Ronald Coase, and centres around the idea that, under certain conditions, 'the problems of external effects can be overcome through private negotiation directly between the affected parties regardless of the initial allocation of property rights'.[11] In other words, the principle offers a purely market-based solution to environmental problems in that the beneficiary of the ecosystem service directly pays the provider irrespective of the initial allocation of property rights, using private money on a voluntary basis following private negotiations.

---

9 Walters Nsoh and Colin T. Reid, 'Privatisation of biodiversity: who can sell ecosystem services?'(2013) 25 *Environmental Law and Management* 12, 14–17.

10 Claudia Sattler and Bettina Matzdorf, 'PES in a nutshell: from definitions and origins to PES in practice – approaches, design process and innovative aspects' (2013) 6 *Ecosystem Services* 2.

11 *Ibid.* 3; Stefanie Engel, Stefano Pagiola and Sven Wunder, 'Designing payments for environmental services in theory and practice: an overview of the issues' (2008) 65 *Ecological Economics* 663, 665.

The Pigouvian type PES follows on from the Coasean principle but recognises that certain types of governmental interferences in the purely market-based approach can also be considered to meet the PES criteria. In the Pigouvian type PES, the government therefore intervenes by either providing payment itself or instigating others to pay on behalf of the direct beneficiaries in order to incentivise ecosystem service provision. In this case, the agreement between the parties is not necessarily a matter of voluntary private negotiation, but it can be driven by the need to comply with regulation, both on the demand and supply side. Evidence from the literature suggests that while there remains some conceptual uncertainty about the meaning of the term,[12] most existing definitions of PES would generally fall between the Coasean and Pigouvian principles (see Box 3.1).

Whatever the label or categorisation, PES is simply a mechanism whereby payments (in cash or in kind) are provided in exchange for the management of land to maintain or enhance the health of the ecosystem, thereby providing for the public or specific beneficiaries benefits like carbon storage or control of water resources. Put simply, it is about paying for benefits (services) from the land, but more specifically linking the suppliers of benefits, such as upstream land managers, to those who benefit from them, such as the downstream communities who use water, to the extent that it constitutes a market-based approach to dealing with environmental challenges.[13] Viewed from this perspective, the scheme perceives the providers of ecosystem services just like other service providers who thus should be rewarded for their work. In this regard, ecosystem services are considered as 'regular' economic services, with the only difference from other services being that provision must be done in a manner that protects biodiversity and the service itself.[14]

---

12 For an overview of the definitions of PES, see Sandra Derissen and Uwe Latacz-Lohmann, 'What are PES? A review of definitions and an extension' (2013) 6 *Ecosystem Services* 12.

13 Genevieve Bennett, Nathaniel Carroll and Katherine Hamilton, *Charting New Waters: State of Watershed Payments 2012* (Forest Trends, 2013) 1, available at www.ecosystemmarketplace.com/reports/sowp2012.

14 McAfee, 'The contradictory logic of global ecosystem services markets' (n. 2) 106.

BOX 3.1 SOME DEFINITIONS OF PES[a]

PES describes a variety of innovative, market-based incentive schemes that reward land managers for maintaining and enhancing environmental benefits ('ecosystem services') such as water regulation (e.g. water quality, flood regulation) and climate regulation.[b]

PES provides an opportunity to put a price on previously un-priced ecosystem services like climate regulation, water quality regulation and the provision of habitat for wildlife and, in doing so, brings them into the wider economy.[c]

PES is a term used to describe a range of schemes through which the beneficiaries, or users, of ecosystem services provide payment to the stewards or providers of those services. The beneficiaries may be individuals, communities, businesses or public bodies.[d]

A PES is: (1) a voluntary transaction where (2) a well-defined ecosystem service (ES) (or a land-use likely to secure that service) (3) is being 'bought' by a (minimum one) ES buyer (4) from a (minimum one) ES provider, (5) if and only if the ES provider secures ES provision (conditionality).[e]

PES refers to the variety of arrangements through which the beneficiaries of ES pay back the providers of those services to ensure their sustainability and timely provision.[f]

[PES] generally have two common features. First, they are voluntary. Second, participation involves a contract between the conservation agent and the landowner. The landowner agrees to manage an ecosystem according to agreed-upon rules and receives a payment (in-kind or cash) conditional on compliance with the contract.[g]

PES schemes rely on incentives to induce behavioural change and can thus be considered part of the broader class of incentive- or market-based mechanisms for environmental policy.[h]

[PES are conceptualised as] new institutions designed to enhance or change natural resource managers' behaviour in relation to ecosystem management through the provision of economic incentives.[i]

[PES are] a transfer of resources between social actors, which aims to create incentives to align individual and/or collective land use decisions with the social interest in the management of natural resources.[j]

Notes: [a] Adopted from Claudia Sattler and Bettina Matzdorf, 'PES in a nutshell: from definitions and origins to PES in practice – approaches, design process and innovative aspects' (2013) 6 *Ecosystem Services* 2, 3.
[b] DEFRA, *Defra Payments for Ecosystem Services (PES) Pilot Projects: Review of Key Findings of Rounds 1 and 2, 2011–2013* (DEFRA, 2014) 4.
[c] S. Smith and others, *Payments for Ecosystem Services: A Best Practice Guide* (DEFRA, 2013) 14.
[d] Ecosystems Knowledge Network, *Payments for Ecosystem Services* (2015), available at http://ecosystemsknowledge.net/resources/tools-guidelines/pes/.
[e] Sven Wunder, *Payments for Environmental Services: Some Nuts and Bolts* , CIFOR Occasional Paper 42(1) (2005) 3.
[f] World Wide Fund for Nature (WWF), *Ecosystem Services and Payments for Ecosystem Services: Why Should Businesses Care?* (WWF, 2007) 4, available at http://assets.panda.org/downloads/business_brochure_1.pdf.
[g] Paul J. Ferraro, 'Asymmetric information and contract design for payments for environmental services' (2008) 65 *Ecological Economics* 810.

[h] B. Kelsey Jack, Carolyn Kousky and Katharine R.E. Sims, 'Designing payments for ecosystem services: lessons from previous experience with incentive-based mechanisms' (2008) 105 *Proceedings of the National Academy of Science of the USA* 9465.
[i] Roger Claassen, Andrea Cattaneo and Robert Johansson, 'Cost-effective design of agri-environmental payment programs: US experience in theory and practice' (2008) 65 *Ecological Economics* 737, 745.
[j] Roldan Muradian and others, 'Reconciling theory and practice: an alternative conceptual framework for understanding payments for environmental services' (2010) 69 *Ecological Economics* 1202, 1205.

Matching those who benefit from a service with those who provide it can be difficult and may be best done through grant and subsidy (public payment) schemes, although more individualised arrangements (private payment or public-private payment schemes) are possible, for example, a company paying the landowner who is providing it with a biodiversity or carbon offset. It should therefore be recognised that as well as a number of documented PES schemes being in operation, or proposed, around the world,[15] the underlying idea features in a number of other regimes. Public payment schemes are not entirely new – historically the public has paid for some of the non-market benefits of ecosystem services through payments to farmers from schemes such as agri-environment schemes under the EU Common Agricultural Policy (CAP) – but PES can be used to seek better targeting and value for money from these existing funding streams, by providing payments for specified results or outcomes.[16]

15 According to the Ecosystem Marketplace's *State of Watersheds Payments* 2012 report, watershed payments totalled more than US$8 billion in 2011 and with evidence of a substantial step up in the number of new initiatives that protect and restore forests, wetlands, and other water-rich ecosystems in 2012: Bennett, Carroll and Hamilton, *Charting New Waters* (n. 13) ii. Payments for watershed services seek to link upstream land use and management with downstream water use and management to realise benefits for upstream and downstream participants in the scheme and for the wider environment.

16 Compliance with environmental standards is now a requirement of any payments within EU agricultural policy (Council Regulation (EC) 1782/2003 of 29 September 2003 establishing common rules for direct support schemes under the common agricultural policy and establishing certain support schemes for farmers and amending Regulations (EEC) 2019/93, (EC) 1452/2001, (EC) 1453/2001, (EC) 1454/2001, (EC) 1868/94, (EC) 1251/1999, (EC) 1254/1999, (EC) 1673/2000, (EEC) 2358/71 and (EC) 2529/2001, [2003] OJ L27/1, arts 3–5). There is also evidence that some payments to farmers from the agri-environment schemes are linked directly to outcomes, e.g. farmers on Islay and other parts of Scotland are paid for lost productivity according to the numbers of geese which feed in their fields during the winter; Scottish Natural Heritage,

Again this is not new, with management agreements based on some payment to landowners in exchange for the land being managed with a view to biodiversity forming a mainstay of British conservation policy, which for decades eschewed more intrusive controls.[17] PES, though, is also seen as a potential way of finding new and additional sources of funding to provide a range of benefits, such as biodiversity conservation, for which there is currently no 'market' or funding. Here, a greater role is envisaged for the private sector with scope for private funding, relieving public bodies of (some) monitoring and enforcement duties. This means that nature conservation would no longer be wholly subject to an official body and its resources, especially at a time of reduced public finances.

Providing funding for services is important since it can be less expensive to pay regularly for ecosystem services provision and maintenance than trying to replace these services with costly industrial infrastructural alternatives once they have been exhausted. For example, it has been estimated that the total value of pollination services to UK agriculture was £603 million in 2010 (although pollination of wild plants also supports a range of other ecosystem services such as carbon sequestration, soil, water quality, and biodiversity, which is likely to have a very high, but currently unknown, market value), and the cost of replacing insect pollination with artificial means would be about £1.9 billion,[18] making artificial replacement a prohibitively expensive alternative.[19] Meanwhile, it is also estimated that the public would be willing to pay between £1.3–1.8 billion per year to conserve pollinators.[20] This suggests that a scheme of payments for pollination services presents the most viable option for conserving insect pollinators, rather than allowing

---

Islay Sustainable Goose Management Strategy October 2014–April 2024 (2014), available at www.snh.gov.uk/docs/A1434517.pdf.

[17] Christopher P. Rodgers, *The Law of Nature Conservation* (OUP, 2013) 112–21.

[18] Environmental Audit Committee, *Pollinators and Pesticides (Seventh Report of 2012–13)*, HC 668 (2012–13) para. 94.

[19] Other estimates reveal that if bees did not pollinate apples and producers had to rely on hand pollination, the price of dessert apples would rise by about 120 per cent if production were maintained at existing levels; *ibid.* para. 94. See further Parliamentary Office of Science and Technology, *Insect Pollination*, POSTnote 348 (January 2010) 2.

[20] Written evidence submitted by Prof. Simon Potts to Environmental Audit Committee (n. 18) Ev. 233.

their continuous decline through business as usual[21] while attempting to replace them with artificial alternatives. A similar calculation to avoid the need for costly replacement inspired the New York City Catskill Watershed PES scheme which was organised to protect and to ensure that the city continue to enjoy high quality, affordable drinking water. To meet the drinking water needs of its residents, the City needed a purification plant large enough to clean its water supply at a cost estimated at between US$8–10 billion to construct the facility and approximately US$1 million each day to operate and maintain the plant.[22] Conversely, the cost of preserving the watershed was estimated at US$1.5 billion,[23] which offered a viable alternative to meeting the drinking water needs of the city's residents without spending the estimated amounts on a purification plant.

The fact that in both these examples the role of PES in conservation is secondary, coming into play only if the replacement alternative to the ecosystem service is prohibitively expensive, is important. Although the mechanism is largely driven by selfish economic considerations, which may provide the major motivation not just for public bodies but also for the private sector, their participation could nonetheless lead to incidental biodiversity benefits. In both examples, the conservation of the ecosystem services could be achieved through public payment schemes,[24] although it would be a challenge to find the resources to pay for a range of biodiversity conservation benefits for which there is currently no 'market' or funding without looking for new and additional sources, involving greater reliance on the private sector. But as with all the mechanisms discussed in this book, the greater the reliance on PES to deliver effective conservation, and the greater the reliance on the private sector within that

---

[21] Some of the factors contributing to the decline in insect pollinators are agricultural intensification, pests and diseases, agrochemicals (insecticides, pesticides), urbanisation, climate change and a reduction in beekeeping (which in Britain is dominated by amateurs); Parliamentary Office of Science and Technology, *Insect Pollination* (n. 19) 2–3.

[22] New York State, *Facts about the NYC Watershed* (2015), available at www.dec.ny.gov/lands/58524.html.

[23] Alice Kenny, *Ecosystem Services in the New York City Watershed* (Ecosystem Marketplace, 2006), available at www.ecosystemmarketplace.com/articles/ecosystem-services-in-the-new-york-city-watershed-1969-12-31/.

[24] In fact, it has been suggested that one way for mitigating the impact of insecticides on pollinators, is the adoption of landscape management approaches in the agri-environment schemes, which could provide a number of benefits to insect pollinators. Written evidence submitted by Prof. Simon Potts (n. 20) Ev. 235.

approach, the more exact, robust and secure the mechanism must be. Establishing the sought after 'exact, robust and secure' mechanism necessary to inspire confidence in the scheme is not an easy task.

Unlike biodiversity offsetting, which is seen to incorporate an element of the 'polluter pays principle' since developers who damage a habitat pay for the provision of compensatory habitat expansion or restoration elsewhere,[25] the focus of PES is on the 'beneficiary pays principle' which creates a dilemma when industrial actors are involved as service providers in PES.[26] While the beneficiary pays principle is considered an exception to the polluter pays principle, especially in cases when polluters are not solvent and where there is no appropriate regulations in place to control the activities of the polluter,[27] it has also been argued that if implemented on a large scale, it can risk deterring reforms in environmental practices and might actually reverse the polluter pays principle.[28] As Aerni has noted in the context of agriculture:

> A pro-poor PES approach will represent a further shift from the polluter pays to a 'pay the polluter principle' because semi-subsistence farmers in Africa that are meant to be the providers of environmental services are often part of the growing environmental problems (deforestation, cultivation in riparian areas, soil impoverishment, and soil erosion owing to a lack of soil replenishment).[29]

If ignored, such a perverse outcome risks undermining the very basis of PES, that is, to ensure that those who provide (or do not destroy) ecosystem services are paid for doing so, saving the vast cost of replacing them by providing a financial incentive to ensure their continued provision. Yet dealing with such risks may require a flexible approach to implementing PES which, on the one hand, could involve

---

[25] See Chapter 4.

[26] Romain Pirard, Raphaël Billé and Thomas Sembrés, 'Upscaling payments for environmental services (PES): critical issues' (2010) 3 *Tropical Conservation Science* 249, 251–3.

[27] *Ibid.* 253; Volker Mauerhofer, Klaus Hubacek and Alastor Coleby, 'From polluter pays to provider gets: distribution of rights and costs under payments for ecosystem services' (2013) 18 *Ecology and Society* 41.

[28] Romain Pirard, Raphaël Billé and Thomas Sembrés, *Questioning the Theory of Payments for Ecosystem Services (PES) in Light of Emerging Experience and Plausible Developments*, Studies No. 04/2010 (IDDRI, 2010) 1.

[29] Philipp Aerni, *The Sustainable Provision of Environmental Services: From Regulation to Innovation* (Spinger, 2015) 38–9.

stakeholders entering into transactions on a voluntary basis,[30] while on the other hand, may involve mandatory elements, driven by compliance regulation, both on the demand and the supply side.

Whatever the approach, the role to be played by PES in support of nature conservation is important. A strong private sector market-based approach with fewer restrictions might call for the widespread use of PES in place of restrictive conservation measures devoted to particular sites or features. Evidence from existing schemes shows that, at the very least, a functional PES scheme should involve a well-defined ecosystem service (or a land-use likely to secure that service);[31] a (minimum of one) service buyer; a (minimum of one) service provider, with payments subject to the service provider securing the ecosystem service provision (conditionality).[32] Others, though, would challenge the view underlying these requirements, on the basis that because elements of biodiversity are not readily quantifiable, what is often paid for is not the service delivered.[33] Although such a simplistic characterisation of PES schemes may be imperfect, it may be necessary, given the range of difficult technical and legal questions surrounding the mechanism, to start at such a simplistic level and build up if we are to harness the potential in the approach to support and improve on the current system of nature conservation. Some would disagree with this approach of building up from the simple model, and the rest of this chapter focuses on how PES schemes work and some of the fundamental design issues that would have to be reconsidered if the mechanism is to be widely adopted at various levels to successfully support and improve on the current system of nature conservation.

---

30 Sven Wunder, *Payments for Environmental Services: Some Nuts and Bolts*, CIFOR Occasional Paper 42(1) (2005) 3. This is, however, with the view that PES can only be used as a supplement to the existing conservation measures, thereby ensuring that we are not paying for action or inaction that is prohibited under existing measures. 'PES must not be seen or used as a substitute for the "polluter pays principle" established by statutory requirements for resource management': DEFRA, *Developing the Potential for Payments for Ecosystem Services: An Action Plan* (DEFRA, 2013) 12.

31 See section 3.2 below.

32 *Ibid.*; S. Smith and others, *Payments for Ecosystem Services: A Best Practice Guide* (DEFRA, 2013).

33 See section 3.2 below.

3.2 WHAT BIODIVERSITY 'SERVICE' IS BEING PAID FOR?

The type of service being paid for determines what kind of PES mechanism is to be adopted: private or public. Biological diversity, or biodiversity, is defined by the Convention on Biological Diversity as:

> [T]he variability among living organisms from all sources including, *inter alia*, terrestrial, marine and other aquatic ecosystems and the ecological complexes of which they are part; this includes diversity within species, between species and of ecosystems.[34]

As noted in section 1.1 above, the power of the definition is that it covers several different but interconnected forms of diversity, all of which must be considered if the natural heritage of the earth is to be passed down to future generations. This definition includes biological resources that have not been commercialised as well as those that have been. Biological resources, which have often been commercialised, are defined as 'genetic resources, organisms or parts thereof, populations, or any other biotic component of ecosystems with actual or potential use or value for humanity'.[35] Deciding on which PES approach to adopt in economic terms requires mapping the different economic characteristics that define private and public goods and services against recognised biodiversity products and services.[36] A private good or service is both rival in consumption, that is, one person's consumption depletes the good's availability to others, and excludable, that is, it is feasible to exclude people (for example, by charging a price) from consuming the good.[37] This would make these commercialised biological resources private goods or services. A public good or service is neither rival nor excludable, and the benefits of preserving biodiversity (broadly conceived) are perceived to have pure public good characteristics with non-rivalry and non-excludability in consumption.[38]

---

[34] Convention on Biological Diversity, Art. 9.

[35] *Ibid.* Art. 2.

[36] Dan Biller, 'The economics of biodiversity loss' in Bjørn Lomborg (ed.), *Solutions for the World's Biggest Problems* (CUP, 2007); OECD, *Harnessing Markets for Biodiversity: Towards Conservation and Sustainable Use* (OECD, 2003); Geoffrey Heal, *Nature and the Marketplace* (1st edn, Island Press, 2000).

[37] OECD, *Harnessing Markets for Biodiversity* (n. 36) 8.

[38] *Ibid.* 8; OECD, *Providing Agri-environmental Public Goods through Collective Action* (OECD, 2013) 226; Alistair McVittie, Dominic Moran and Steven Thomson, *A Review of Literature on the Value of Public Goods from*

Because of the public goods character of the benefits of preserving biodiversity, if the biodiversity good or service is not one that has yet been 'commercialised' and therefore subject to 'observable market transactions'[39] (for example, food), it may be difficult to see how the voluntary element following private negotiations can properly operate, not just because of the absence of rivalry and exclusion among the beneficiaries, but because the production of such a commercialised good or service will quite often generate biodiversity externalities that fall within the scope of public goods.[40] This highlights the difficulty of disentangling services, both spatially and in terms of the same management activity furthering several services, for example, maintaining a healthy peat bog provides benefits for both biodiversity and carbon storage, whilst contributing with many other elements in the wider catchment to the quality and quantity of water resources. For this reason, the preferred payment scheme may have to involve some kind of government intervention, either by making payments itself or requiring others to pay on behalf of the direct beneficiaries, who are the public.

Irrespective of who the buyer is likely to be, biodiversity itself is not a single quantifiable element, and whether or not a PES scheme enhances the protection that biodiversity receives will depend on how the scheme is designed and the legal framework underpinning it. While some schemes are designed to pay for the provision or enhancement of a single non-biodiversity related service, even these are likely to have incidental benefits for biodiversity,[41] and there is a growing interest in schemes that protect or restore multiple ecosystem services in addition to the targeted services of interest. These schemes either 'stack'[42] multiple individual services, where payments are then made for different ecosystem services

Agriculture and the Production Impacts of the Single Farm Payment Scheme, Report Prepared for the Scottish Government's Rural and Environment Research and Analysis Directorate, RERAD/004/09, SAC (2009) 3–6.

39 *Ibid.* 6.

40 *Ibid.* 6; OECD, *Providing Agri-environmental Public Goods Through Collective Action* (OECD, 2013) 226–7.

41 E.g. the New York City Catskill Watershed PES scheme was designed to provide the city with high quality, affordable drinking water. The provision of the drinking water service had an incidental benefit of preserving the watershed and its biodiversity; Kenny, *Ecosystem Services in the New York City Watershed* (n. 23). Similar biodiversity benefits can be found in other drinking water provision PES schemes around the world; Bennett, Carroll and Hamilton, *Charting New Waters* (n. 13).

42 David Cooley and Lydia Olander, 'Stacking ecosystem services payments: risks and solutions' (2012) 42 *Environmental Law Reports* 10150; Jussi

separately from the same system, or 'bundle'[43] multiple individual services together in a single package to be bought by individual or multiple buyers, thereby realising potentially greater environmental and market efficiency benefits than single service schemes can deliver (see Table 3.1).

On the one hand, with stacking, it may be possible to tailor the marketing of a specific type of service more effectively to specific types of buyers, for example, a water utility company (interested in the water quality service of a watershed), industrial corporations (interested in the carbon storage service of peatland) and land developers (interested in a biodiversity gain on peatland for an offset). On the other hand, bundling multiple individual ecosystem services can allow sellers to charge a premium for the provision of this range of services. For this to happen, though, all the services must be clearly identifiable and quantifiable, in the sense that the seller has knowledge of the existence and value of the service, although it could be argued that one advantage of bundling is that it is not necessary to specify all the services covered, just the main one(s) and 'the associated ones'. Bundling ecosystem services has the potential to reduce the transaction costs associated with establishing a market for multiple individual services, both by reducing the number of individual agreements that may have to be made between the beneficiaries and providers for each individual service and by reducing the scope for disagreements (for example, of valuation) which can make it difficult to bring beneficiaries and providers together.[44] However, the possibility of charging a premium for the bundled services may not be very attractive to policy-makers and potential buyers of a particular service, for example, a water company interested only in water quality.

Despite the potential benefits, both stacking (also known as 'layering') and bundling present challenges in terms of programme design and implementation. Given the difficulty in identifying biodiversity as a single quantifiable unit, it may therefore be the case that effective conservation of biodiversity using PES schemes could be achieved

---

Lankoski and others, *Environmental Co-benefits and Stacking in Environmental Markets*, OECD Food, Agriculture and Fisheries Papers No. 72 (OECD, 2015).

[43] Sven Wunder and Sheila Wertz-Kanounnikoff, 'Payments for ecosystem services: a new way of conserving biodiversity in forests' (2009) 28 *Journal of Sustainable Forestry* 576; Robert L. Deal, Bobby Cochran and Gina LaRocco, 'Bundling of ecosystem services to increase forestland value and enhance sustainable forest management' (2012) 17 *Forest Policy and Economics* 69.

[44] See section 3.3 below.

through bundling,[45] but there are issues of governance to contend with.[46] Designers of any scheme would not only have to work out the service or the management intervention that is being paid for in order to ensure that resources are adequately targeted to enhance the conservation of nature, but also to ensure that those who receive payments have genuinely contributed to the specific service provision.

At the extreme end of the spectrum is the option known as 'piggy-backing' where not all of the ecosystem services generated from a single parcel of land or body of water are valued and sold to buyers. Instead, the buyer identifies a single service of interest, which is bought as an umbrella service, whilst the buyer/beneficiaries benefit free of charge from the other services provided by the land. For example, a developer pays an upstream land manager for peatland restoration to reduce/offset net greenhouse gas emissions. Maintaining the peatland simultaneously provides habitat for wildlife which may be counted as a biodiversity gain for an offset, improves water quality and enhances recreational values.[47] However, no buyers are found for these additional services and the benefits they provide are received at no cost to end-users. Again, the challenge with choosing this option is with how to prevent those who have not contributed to the provision of the services from benefiting (free-riding) or how to get them to pay. The problem of free-riders is most easily seen in the context of biodiversity conservation under the present position where everybody free-rides on the landowners who are bearing the burden of restrictions on the use of their land but usually receiving no reward for the services provided. This is likely to continue to be the case, especially where there is no link to an offsetting or some other regulatory requirement for the buyer. There is also a danger that it may facilitate the engineering of ecosystems to maximise single service provision and revenue flows, with potential collateral damage to other aspects of biodiversity.[48]

---

[45] In fact, studies reveal that schemes in China and the United States where some stacking or bundling has been reported show that the majority of bundling involves biodiversity services: Bennett, Carroll and Hamilton, *Charting New Waters* (n. 13) 12.

[46] See section 3.6 below.

[47] M.S. Reed and others, *Peatland Code Research Project Final Report* (DEFRA, 2013).

[48] E.g. the New York City and Catskills Watershed scheme was driven by the economically attractive decision for the city to invest in the provision of the single ecosystem service of water purification which was justified by the avoided cost of not building a water treatment plant. For more on this see Gretchen C.

Table 3.1 Different approaches to 'packaging' ecosystem services

A. Bundling A single buyer, or consortium of buyers, pays for the full package of ecosystem services that arise from the same habitat.	
Climate regulation Local flood risk reduction Landscape quality Water quality Habitat for wildlife	Single buyer for all services
B. Layering (or stacking) Multiple buyers pay for the separate ecosystem services that are supplied by a single habitat.	
Climate regulation	Individual buyer
Local flood risk reduction	Individual buyer
Landscape quality	Individual buyer
Water quality	Individual buyer
Habitat for wildlife	Individual buyer
C. Piggy-backing Not all of the ecosystem services produced from a single habitat are sold to buyers. One (or a few) service(s) is sold as an umbrella service, whilst other services are said to 'free-ride', i.e. the benefits they provide are received by users free of charge.	
Climate regulation	Buyer
Local flood risk reduction	Free-riding
Landscape quality	Free-riding
Water quality	Free-riding
Habitat for wildlife	Free-riding

Source: Adapted from S. Smith and others, *Payments for Ecosystem Services: A Best Practice Guide* (DEFRA, 2013) 22.

The United Kingdom has started a number of pilots with the aim to develop institutional arrangements to address some of the issues of costs,

Daily and Katherine Ellison, *The New Economy of Nature: The Quest to Make Conservation Profitable* (Island Press, 2002) 61–85; James Salzman, 'Creating markets for ecosystem services: notes from the field' (2005) 80 *New York University Law Review* 870, 889–92.

including the potential to develop bundled or stacked PES schemes,[49] and around the world, research has identified a number of programmes that engage in some form of stacking or bundling.[50] According to the Ecosystem Marketplace *State of Watersheds Payments* 2012 report, almost a third of schemes from around the world identified in the study involved some form of bundling and stacking, with the conservation of habitats that support biodiversity the most common ecosystem service targeted.[51] How the bundled ecosystem services are quantified, protected or monitored varies across programmes, but in relation to biodiversity, it would seem that the schemes operate on the assumption that habitat protection will deliver biodiversity benefits to the extent that there is no need to engage in a wholesale quantification of each individual service to be paid for. Identification of the multiple individual services may also be possible through detailed tracking and the management strategies adopted to enhance or support the provision of such services. Yet mere identification of the individual services may not be sufficient for developing a credible PES scheme that can attract a significant number of providers to enable a functioning market – the market value of these services still has to be determined.

3.3 VALUATION OF ECOSYSTEM SERVICES

An important aspect of designing a credible PES scheme is how the services to be paid for are valued. Valuation is important not just in determining the price to be paid for each ecosystem service provided, but it can also provide a tool by which the economic loss resulting from the destruction of ecosystem services is no longer viewed separately from the economic gain of development.[52] Yet attaching a value to ecosystem services in itself is not a guarantee for the sustainable management of services. Although valuation can play an essential role in creating markets for the conservation of biodiversity and ecosystem services,[53] on

---

[49] See e.g., DEFRA, *Defra Payments for Ecosystem Services (PES) Pilot Projects: Review of Key Findings of Rounds 1 and 2, 2011–2013* (DEFRA, 2014).

[50] Bennett, Carroll and Hamilton, *Charting New Waters* (n. 13) vii.

[51] *Ibid.* 10.

[52] The Economics of Ecosystems and Biodiversity (TEEB), *The Economics of Ecosystems and Biodiversity* (n. 8) 3.

[53] Engel, Pagiola and Wunder, 'Designing payments for environmental services in theory and practice' (n. 11); Unai Pascual and others, 'Special

the one hand, putting a value on the ecosystem services lost is likely to be seen by those who manage land as adding further environmental restrictions to development of land. This not only affects the decision-making process but also exposes the decisions on values and the effect of proposed changes to management practices to challenge by land managers who may not agree with them.[54] On the other hand, any charge takes the service provision out of the category of what is normally expected, so that the landowner can now receive (and demand) a reward for the services, which were previously provided free of charge.[55] Viable schemes of reward must, therefore, find a way to exclude the free-riders from benefiting from the charged service by getting them to pay the full 'market price' for the service.

An approach that has been suggested to ensure that decisions about economic development do not overlook or undervalue the loss (economic and in other terms) of ecosystem services is the emphasis on linking the costs and benefits more directly.[56] Even so, complications arise when attempting to determine the value of the costs and benefits because specific non-marketable, but nonetheless valuable, ecosystem services such as those derived from biodiversity, cannot readily be priced.[57] Their relative or absolute value is implicit rather than explicit, while other ecosystem services such as producing food, timber and energy do have an explicit economic value reflected in their market price. The consequence of this is that under most PES schemes, including the agri-environment schemes under EU legislation, the ecosystem services derived from biodiversity are often not valued financially in their own right. With a few exceptions, they are rather assessed by reference to the

---

section: payments for environmental services: reconciling theory and practice' (2010) 69 *Ecological Economics* 1202. Although the 'market-based' label is used for PES, it has also been observed correctly that 'not all payments are markets', with a market defined as 'a constellation of buyers and sellers involved in transactions': R Muradian and others, 'Payments for ecosystem services and the fatal attraction of win-win solutions' (2013) 6 *Conservation Letters* 274, 276.

54 Nsoh and Reid, 'Privatisation of biodiversity' (n. 9) 14–15.

55 Muradian and others, 'Payments for ecosystem services' (n. 53) 276–7.

56 DEFRA, *An Introductory Guide to Valuing Ecosystem Services* (DEFRA, 2007). See also Colin T. Reid, 'The privatisation of biodiversity? Possible new approaches to nature conservation law in the UK' (2011) 23 *Journal of Environmental Law* 203, 220–22.

57 See Colin T. Reid 'Between priceless and worthless: challenges in using market mechanisms for conserving biodiversity' (2013) 2 *Transactional Environmental Law* 217; Dieter Helm and Cameron Hepburn 'The economic analysis of biodiversity: an assessment' (2012) 28 *Oxford Review of Economic Policy* 1.

land area or income forgone by adopting specified land management practices or possibly costs incurred, for example, for planting hedges or creating ponds. In other words, the service being paid for has no direct link to the means by which the payment is calculated.

While acreage might be appropriate in calculating payments in schemes where bundling is the preferred option and for determining the value of some ecosystem services (such as protecting soil productivity that mainly requires leaving land in its natural state for some time), it cannot provide an adequate basis for valuing other services from land in their own right, and does not ensure that payments go to the person directly or indirectly responsible for the provision of a particular service over time. This is a fundamental design issue with input-based PES schemes, as the system of payment rewards the present land manager's efforts for implementing changes needed for the provision of services, but without tying payments to service output. It is therefore difficult to ascertain if payments have been made to the right person, especially for services such as those derived from woodland, where it clearly takes beyond the lifetime of the current landowner or occupier for trees to reach any appreciable size and may take several centuries before there is a varied woodland community, with a mix of young, mature, senescent and dead trees that can provide the range of ecological niches required to fulfil the biodiversity potential of the site.

The task for the designers of any PES scheme, then, is to ensure that there is a balance between the rates of payment, the service being provided and those receiving payments at a given time. This is an issue that is evident in the US Conservation Reserve Program (CRP). Under the scheme, annual payments are provided to farmers for planting long-term, resource-conserving covers on eligible farmland.[58] The CRP encourages farmers to convert highly erodible cropland to vegetative covers, such as native grasses, wildlife plantings, trees, filter strips or riparian buffers. In so doing, the farmers are paid for the provision of ecosystem services such as protection against soil erosion, preservation of water quality and provision of wildlife habitat. Payment rates are primarily based on relative productivity of the soils within each county

---

[58] See generally the website of the Conservation Reserve Program (CRP), www.fsa.usda.gov/programs-and-services/conservation-programs/conservation-reserve-program/index. For an overview of the history and development of the CRP, see J.B. Ruhl, Steven E. Kraft and Christopher L. Lant, *The Law and Policy of Ecosystem Services* (Island Press, 2007) 186–92.

and the average dry land cash rent or cash rent equivalent;[59] determining the amount to pay is still largely dependent simply on the area of land to be placed under the scheme. Moreover, in addition to the payment rates which are calculated in advance of the land delivering any service, the scheme also pays up to 50 per cent of the cost incurred in the process of establishing the required management changes on the land.[60]

Although there have been reported benefits, such as reduced soil erosion, improved water and air quality, increased wildlife habitat and populations, and wider societal benefits in the 30 years of the CRP,[61] the challenge remains to ensure that there is a balance between the rates of payment, the service being provided and those receiving payments at a given time. At the moment, a broad interpretation of the land and producer eligibility criteria under the CRP (see Box 3.2), reveals that the programme may allow for the enrolment of lands and for payments to be made to those that do not contribute significantly to service provision, while farmers who have taken good care of their lands, and so do not need to implement changes to deliver substantial services, may not qualify for payments.[62] With such broad eligibility criteria there is a risk of reward to people who destroy and then restore habitat,[63] but also the

---

[59] There is a bonus of up to 20 per cent of the agreed annual rental payments for some land of particular environmental value as an incentive to perform certain maintenance obligations; USDA Farm Services Agency (FSA), *Prospective Participants/General Public* (2015), available at www.fsa.usda.gov/programs-and-services/conservation-programs/prospective-participants/index; Ruhl, Kraft and Lant, *The Law and Policy of Ecosystem Services* (n 58) 190–91.

[60] USDA Farm Services Agency (FSA), *Prospective Participants/General Public* (n. 59).

[61] For some of the success stories of the CRP across the United States, see the interacting website, USDA Farm Services Agency (FSA), *Celebrating 30 Years of the Conservation Reserve Program* (2015), available at http://usdaonline.maps.arcgis.com/apps/MapTour/index.html?appid=588d05579d3045479e36d2abf6ab5456.

[62] USDA Farm Services Agency (FSA), *Prospective Participants/General Public* (n. 59); Salzman, 'Creating markets for ecosystem services' (n. 48) 894.

[63] E.g. in the Poole Harbour Catchment Nitrogen PES Scheme (a DEFRA pilot project), one of the main barriers to a scheme paying farmers for ongoing land management changes to reduce nitrogen levels and improve water quality linked to nutrient offsetting of new development was the reluctance among local authorities for any use of developers' funds to address agricultural pollution. This was seen as unfair and politically unacceptable given that new development is expected to comprise less than 2 per cent of total nitrogen load from the catchment while agriculture contributes 85 per cent: RSPB, *The Feasibility of a Nitrogen PES Scheme in the Poole Harbour Catchment* (DEFRA, September

question of how to factor in to the valuation process the risk of a breakdown of the ecosystem service, for example, from climate change. These are largely ethical issues which are dealt with further in section 3.7 below, but in practice and depending on why the services are 'required' or what the buyer's motivation/incentive is, an input-based PES scheme will only thrive if the buyer is satisfied that the agreed management practices or capital projects will indeed deliver the required ecosystem services, further underlining the voluntary nature of PES.[64] From a practical and financial viewpoint, an input-based approach seems like a more logical option than attempting to contract for the delivery of a prescribed level of ecosystem services which may be impractical or unacceptable to the parties, given the unforeseen risks that can have a negative impact on service provision.

BOX 3.2 CRP PRODUCER AND LAND ELIGIBILITY CRITERIA

Producer Eligibility

To be eligible for CRP enrolment, a farmer must have owned or operated the land for at least 12 months prior to the previous CRP sign-up period. Exceptions to this rule include:

- land acquired by the new owner due to the previous owner's death;
- change in ownership due to foreclosure; or
- land that was purchased by the new owner without the sole intention of placing it in CRP.

Land Eligibility

To be eligible for placement in the CRP, land must be either: cropland (including field margins) which is planted to an agricultural commodity for four of the previous six crop years from 2008 to 2013, and which is physically and legally capable of being planted in a normal manner to an agricultural commodity; or certain marginal pastureland that is suitable for use as a riparian buffer or for similar water quality purposes.

2013) 3, available at http://randd.defra.gov.uk/Default.aspx?Menu=Menu&Module=More&Location=None&ProjectID=18082.

64 James M. Gibbons and others, 'Should payments for biodiversity conservation be based on action or results?' (2011) 48 *Journal of Applied Ecology* 1218.

The land must be eligible and suitable for any of the following conservation practices:

> buffers for wildlife habitat; wetlands buffer; riparian buffer; wetland restoration; filter strips; grass waterways; shelter belts; living snow fences; contour grass; strips; salt tolerant vegetation; shallow water areas for wildlife.

Source: USDA Farm Services Agency (FSA), *Prospective Participants/General Public* (2015), available at www.fsa.usda.gov/programs-and-services/conservation-programs/prospective-participants/index.

The simplicity and reduced costs involved explain why in practice input-based PES schemes are more common, even though these may not guarantee that payments are going to the right people.[65] There is also the issue of time-scales, since neither the public nor private sector tends to favour making financial commitments too far into the future, and landowners similarly tend not to like tying up their land. Yet there has been some innovation in relation to output-based payments. For example, in the State of North Carolina in the United States, the Tar-Pamlico river basin water quality was at risk through eutrophication. In response, the 16 regional companies discharging waste water into the river basin made a voluntary agreement to reduce nutrient loads in their waste water. This was done through monitoring and measurement of the nutrients and whenever the defined threshold value for phosphorus and nitrogen was transgressed, they paid a per kilogram fee into the North Carolina Agricultural Cost Share Program, which was used to finance offset projects implemented by local farmers.[66]

The success of the Tar-Pamlico project[67] demonstrates that output-based PES schemes can deliver ecosystem services benefits, but without running the risk of paying people who may not have contributed to the

[65] *Ibid.*

[66] EPA, *North Carolina: Tar-Pamlico Basin – Agricultural Management Strategy Reduces Instream Nutrients* (2005), available at http://water.epa.gov/polwaste/nps/success319/nc_tar.cfm.

[67] Through implementation of best management practices (BMPs) on agricultural lands, such as riparian buffer protection, reduced fertilizer use, and implementation of conservation tillage practices, North Carolina met its 30 per cent nitrogen reduction goal ahead of schedule and impaired acreage in the estuary was reduced by 90 per cent, allowing one section of the estuary to be removed from the 303(d) list for chlorophyll (although the success of the output-based payment method may be down to the relatively easy way of measuring the service); *ibid.*

delivery of the service. Despite the fears of impracticality in measurement and potential for increasing transaction costs, output-based payments can deliver benefits that are acceptable to providers,[68] for example, by creating incentives for managers to use their private knowledge to optimise their level of action and join the scheme only if they can produce the biodiversity services targeted by the scheme.[69] But effective PES schemes do still require well-defined land or resource-tenure regimes for providers.

BOX 3.3 DISTINCTION BETWEEN 'INPUT-BASED' AND 'OUTPUT-BASED' PAYMENTS

- 'Input-based' payments are made on the basis of certain land or resource management practices being implemented. For example, payments might be made for the creation and maintenance of buffer strips along watercourses or the restoration and upkeep of green spaces in residential areas or for planting a hedge or pond-creation where there are direct costs to be met.
- 'Output-based' payments are made on the basis of actual ecosystem services provided. For example, payments might be made for a certain level of carbon sequestration or a measured increase in biodiversity.

Source: S. Smith and others, *Payments for Ecosystem Services: A Best Practice Guide* (DEFRA, 2013) 20.

3.4 EFFECTIVENESS OF MARKETS: WHO CAN SELL ECOSYSTEM SERVICES?

Besides identifying the value of the natural environment to society, the more holistic approach that the ecosystem framework encourages requires a shift in the both the mind-set and practices of many of those who manage and use land. In this commercial/regulatory context farmers,

[68] Another example of a scheme that pays by results on an output basis is the geese management schemes, which pays compensation based on geese numbers to farmers for the lost productivity caused by geese which feed in their fields during the winter in parts of Scotland; Scottish Natural Heritage, *Islay Sustainable Goose Management Strategy October 2014–April 2024* (2014), available at www.snh.gov.uk/docs/A1434517.pdf.

[69] Gibbons and others, 'Should payments for biodiversity conservation be based on action or results?' (n. 64).

for example, are viewed as 'integrated land managers' who produce food and provide ecosystem services rather than merely 'food producers'. If farmers and other land users are now to be seen as integrated land managers, it would seem obvious that those who can receive payments for ecosystem services should be the land managers and users who have the requisite control over what happens on the land. However, fitting responsibility for ecosystem services into land tenure and related structures designed for traditional agriculture and other uses based on providing something other than ecosystem services poses a range of difficulties. In fact, the lack of tenure clarity and security among ecosystem service providers has been identified as perhaps one of the biggest barriers to establishing working PES schemes.[70]

Almost any environmental conservation programme will be improved by a clearly defined land and resource tenure system, to the extent that PES schemes simply cannot operate without it, as there is a need for clear identification of rights and responsibilities associated with service provision on a given piece of land. Indeed, it is on this basis that the ideas of stewardship, which call for a qualification of private property rights, have been built.[71] Historically, the law recognises private ownership of land (and plants as accretions to the soil) but not of wild creatures, and most ecosystem services are not reflected in legally recognised entitlements. For example, although bee-keepers can and do sell the services of the bees they own, there is no legal basis which entitles landowners to charge neighbours for the pollination services provided by wild insects which cross onto the neighbours' land nor which entitles the neighbours to demand these services.

The extensive rights of the owner over the land provide a useful starting point for considering how legal relationships fit with new approaches to conservation. Conventional legal structures can be utilised to the extent that PES goals can be secured simply by restricting the owner's freedom of action, for example, by prohibiting building or other forms of development which will destroy habitat.[72] But where the delivery of the PES goals also requires positive action on the part of the

---

[70] Sven Wunder, 'When payments for environmental services will work for conservation' (2013) 6 *Conservation Letters* 230, 234.

[71] William N.R. Lucy and Catherine Mitchell, 'Replacing private property: the case for stewardship' (1996) 55 *Cambridge Law Journal* 566. For a detailed analysis of different concepts of stewardship, see Emily Barritt, 'Conceptualisng stewardship in environmental law' (2014) 26 *Journal of Environmental Law* 1.

[72] For an analysis of how more positive conservation obligations lead to a new characterisation of property rights see Christopher Rodgers, 'Nature's place?

landowner or occupier, especially to conserve the natural environment beyond the hotspots that are the focus of policy dedicated to protecting the most threatened aspects of biodiversity, the ideals of shared responsibility[73] and stewardship are brought into play. The involvement of more parties through shared responsibility and stewardship offers opportunities for a greater integration between conservation and other aspects of activity and policy but there is a need for clear identification of rights and responsibilities associated with service provision on a given piece of land. If there are no clearly defined rights and responsibilities, there may be overlap between the rights and responsibilities of local land managers who will in turn lack the right to exclude access or other actions by third parties which will affect the provision of the service. Without the ability to exclude other interventions which can have an impact on the features of the land, potential service providers will be unreliable since they lack sufficient control over the delivery of the service and will lack the incentive to continue providing the service. Identifying exactly who is responsible and therefore entitled to payment in order to negotiate with them can also have a significant impact on the transaction costs of operating the PES scheme.[74]

It has been suggested that such clarity of tenure can sometimes be achieved with external help, as part of a collective PES deal,[75] by bringing together all the people who might claim a legal interest in the ecosystem (which might include neighbours, or where there are clearly defined rights that are separated, for example, the landowner and the holder of game rights). Such an approach may be possible in circumstances where community ownership exists or where the nature and limits of each individual's rights are known to a certain extent. However, many ecosystem service rich settings, in both the developing and developed world, are marked by land disputes, overlapping claims, conflicting rights/responsibilities and even violence,[76] making it difficult for a

---

Property rights, property rules and environmental stewardship' (2009) 68 *Cambridge Law Journal* 550.

[73] Declaration of the United Nations Conference on the Human Environment, Stockholm 1972, UN Doc. A/Conf.48/14/Rev 1 (1973) Preamble, para. 7.

[74] See section 3.5 below.

[75] Meine van Noordwijk and others, 'Payments for environmental services: evolution toward efficient and fair incentives for multifunctional landscapes' (2012) 37 *Annual Review of Environment and Resources* 389.

[76] Wunder, 'When payments for environmental services will work for conservation' (n. 70) 234; Nsoh and Reid, 'Privatisation of biodiversity' (n. 9) 19.

collective PES deal to be agreed. Estimates from Brazil show that only about one-quarter of threatened Amazon forestland would have the clear tenure structure that is suitable for the implementation of PES.[77]

In the United Kingdom, complications arise if areas with multiple occupancies such as common land in England and Wales and crofting land[78] in Scotland are to be used for the provision of ecosystem services. Commons are subject to individual ownership with a wide range of property rights, but the owners' freedom is restricted by recognising the legal validity of the use rights of commoners, so that deciding who should receive payments for ecosystem service provision – those who hold the specific commons use rights that affect the ecosystem service or the owners – would depend on what type of rights they are seen to be (Box 3.4).[79] But the commoners' exercise of their use rights may not always be compatible with the management of the land for service provision. For example, the right to cut turf or peat or timber is not always going to be beneficial to the delivery of some ecosystem services, such as those which may depend on maintaining a healthy peat bog to provide benefits for both biodiversity and carbon storage.

Eligibility for payments may also depend on whether PES entitlements in the commons are seen as a new category of commoners' use rights, or rights that should be attached to existing categories of commoners' rights, or are among the rights retained by the landowner. Given that the provision of services on common land will invariably be dependent at least in part on whether and how other rights are exercised, it is unlikely that the entitlement to payments under any PES scheme on common land could exist exclusively as a new category of commoners' rights.[80]

---

[77] Jan Börner and others, 'Direct conservation payments in the Brazilian Amazon: scope and equity implications' (2010) 69 *Ecological Economics* 1272.

[78] See e.g., Crofters (Scotland) Act 1993, ss. 50, 50A, in relation to using common grazing land for forestry.

[79] For a more detailed analysis of the legal categories of use rights in the commons, see Angela Sydenham, *Commons and Village Greens: The Modern Law* (Lime Legal, 2006) 3.2.1–3.2.7.

[80] Christopher P. Rodgers and others, *Contested Common Land: Environmental Governance Past and Present* (Earthscan, 2011) 163–88; Nsoh and Reid, 'Privatisation of biodiversity' (n. 9) 19–20.

BOX 3.4 LEGAL CATEGORIES OF USE RIGHTS IN THE COMMONS

- **The common of pasture** is the right to graze livestock on the common. The exercise of this right has traditionally been governed by a set of principles that regulate the number of livestock entitled to graze the common resource.[a]
- **The common of turbary** is the right to cut turf or peat for fuel, and is limited to domestic use.
- **The common of estovers** is similar to that of turbary. It is the right to cut timber or other vegetation for purposes of repairing fences, houses or as fuel. It is limited in quantity or limited to what is necessary for the maintenance of the land to which it is attached, and which it is intended to benefit.
- Other rights of common, less commonly encountered in practice today, are the right of pannage (the right to graze pigs in forests), and the right of piscary (the right to take fish from a lake, river or stream belonging to another).

Note: [a] The rule of levancy and couchancy restricted the number of livestock allowed to graze the common based on the capacity of the land and the rule of 'stints' permitted a fixed number of livestock to graze the common land.

Even if clarity and security of tenure can be established through a collective deal as earlier suggested, the fact that different use rights in the commons are not always mutually compatible raises further issues of how responsibilities and benefits accruing from the provision of ecosystem services are shared between different rights holders and the owners. In most of the developing world, particularly in Africa, there are major difficulties over how to reconcile traditional customary land tenure systems that often rely on local knowledge, historical occupation and use, with the western property model that, quite often, is at odds with the customary tenure system.[81] Such lack of clarity as to who owns land has serious implications for who may or may not be entitled to receive payments for the provision of ecosystem services, even though they may

[81] E.g. on the situation in Cameroon, see Walters Nsoh, 'Ecosystem services: a possible new approach in the valuation of compensation for land expropriation in Cameroon' (2013) 4 *IUCNAEL EJournal* 17 (published in revised form as Walters Nsoh, 'Customary land rights, indigenous rights and land expropriation in Cameroon: ecosystem services as a possible new approach in valuing compensation' in Ting Xu and Jean Allain (eds), *Property and Human Rights in a Global Context* (Hart Publishing, 2015).

have contributed to the enhancement of the service over time.[82] A further complication is the fact that ecosystem services will often be provided by areas of land that do not constitute a single specific occupancy unit, for example, depending on all or part of a water catchment, not individual farms or estates. This raises immediate difficulties over identifying who is the provider of a particular service (for example, carbon sequestration) and what share of benefits or burdens is to be borne by a particular plot.

The lack of certainty over tenure is not only restrictive to PES schemes for existing land managers and for developing countries. In most of the developed world where property rules governing land are much more formal, there is also likely to be a mismatch between the need for management over extended periods in order to secure biodiversity benefits and the ever-changing nature of tenure arrangements. This has implications both for the extent to which the owners and occupiers of land can contribute to the provision of ecosystem services and for the social and economic relationships between those who manage land. The task, then, for the designers of any PES scheme is how to ensure that the beneficiaries of the service can ascertain who it is that is responsible and therefore entitled to receive payments for particular service provision in the short, medium and long term. A starting point when using land subject to agricultural tenancy arrangements for the provision of ecosystem services is perhaps knowing precisely what rights (for example, game and minerals rights) the landlord retains. Alternatively, where land is held and managed under a tenancy agreement, as is likely to be the case in much of the British countryside which plays a key role in the provision of ecosystem services, one option is to require the tenant to obtain the landlord's written approval prior to entering any such schemes. This is an approach that has been adopted under the various agri-environment and forestry support schemes, in which the landlord's consent is required before the tenant can proceed to carry out certain activities on the land.[83] Requiring a tenant to obtain the consent of a landlord creates an extra burden; one which has been described in the context of proposed conservation covenants in England as 'an unnecessary extra burden … [which] might create real difficulties for those leaseholders whose freeholder cannot be identified'.[84]

---

82 *Ibid.*

83 Rural Development Contracts (Land Managers Options) (Scotland) Regulations 2008, SSI 2008/159; DEFRA, *Single Payment Scheme Handbook for England 2013 Including the Uplands Transitional Payment* (DEFRA, 2013) 49.

84 Law Commission, *Conservation Covenants*, Law Com. No. 349 (2014) paras 5.17, 5.34.

Yet such consent alone may not be sufficient to safeguard the provision of services in the medium and long term if ownership of the land changes. The challenge then is to work out how payments should be arranged to compensate those responsible for service provision in the short, medium and long terms. Most existing schemes pay for the input, that is, the cost of certain management activity, for example, planting a hedge, but not for number of birds benefiting. This can mean not rewarding those in the medium and long term responsible for maintaining the service flow if ownership or occupation has since changed. One way of providing such safeguards is ensuring that schemes operate within the private law of property, in the same way that conservation covenants would do.[85] It will require that contracts for PES go beyond the privity of contract doctrine where they can bind only the original parties to the agreement, in order to run with the land and therefore be binding on third parties and successors in title. For example, under the US Conservation Reserve Program (CRP), an eligible person may acquire the interest in the CRP as a successor if the land has been sold, there has been a change in owner or operator, or a foreclosure or involuntary loss of land has occurred.[86] The eligible person acquires the land with all CRP entitlements and obligations, which may include any loss in payments as a result of violations by the previous owner.[87] Similarly, in England conservation covenants, which bind land in perpetuity or for any shorter term specified in the covenant,[88] have been promoted as one of the vehicles for delivering PES schemes.[89]

3.5 CONTRACT DESIGN

PES as an approach to enhancing the conservation of nature may involve the obligation to manage land in a particular way for a given period of time to provide ecosystem services in exchange for payments. This means that the parties involved must necessarily enter into an agreement

---

85 See Chapter 5 for a discussion on proposed conservation covenants in England. For details, see Law Commission, *Conservation Covenants: A Consultation Paper*, Law Com. No. 211 (2013) and Law Commission, *Conservation Covenants*, Law Com. No. 349 (2014).

86 USDA Farm Services Agency (FSA), *Prospective Participants/General Public* (n. 59).

87 *Ibid.*

88 Law Commission, *Conservation Covenants* (n. 85) para. 5.44.

89 *Ibid.* paras 2.4–2.55.

of some kind. A binding agreement provides legal certainty by ensuring that parties have the same understanding of their own and the other parties' respective rights, responsibilities and obligations and of where risk is allocated. Such legal clarity as to each party's rights and obligations can help reduce the possibility of dispute and failure later on. This is especially important for services such as biodiversity conservation that may require periods beyond the lifetime of the original parties to the agreement before significant benefits are delivered.

PES contracts can take different forms depending on the kind of specified management practices to be adopted. They could range from a simple document authorising and acknowledging payment for the planting of a hedge to a complex document laying down detailed terms and conditions in relation to the specified management practices to be adopted, monitoring and enforcement; akin to how management agreements are currently used to promote the conservation of nature.[90] The particular PES agreement best suited for nature conservation will therefore depend on the management practices to be adopted as well as the local circumstances. Complex written contracts that are common in developed countries may prove too cumbersome, expensive or not feasible in practice, in some developing and transition countries. This means that depending on the local circumstances, a PES contract may not need to be legally binding to prove effective as long as there is sufficient clarity as to each party's rights and obligations.

Beyond the discussion on the significance of having a contract and the circumstances under which a contract could prove effective, there are many design issues that arise in the development of contracts between a government (or public agency),[91] landowners (or managers) or any

---

[90] On the role of management agreements in practice see Louise Livingstone, Jeremy Rowan-Robinson and Richard Cunningham, *Management Agreements for Nature Conservation in Scotland* (University of Aberdeen, 1990); Christopher Rodgers and Jennifer Bishop, *Management Agreements for Promoting Nature Conservation* (RICS, 1998).

[91] As already discussed in section 3.2 above, because of the public goods character of the benefits of conserving biodiversity, it is likely that some kind of government intervention may be involved, either by providing payments itself or as an intermediary with the public as the buyers. For example, the Forestry Commission (a government agency) acted as an intermediary to private sector PES in developing the Woodland Carbon Code; Forestry Commission, *Woodland Carbon Code: Requirements for Voluntary Carbon Sequestration Projects (Version 1.3)* (Forestry Commission, 2014), available at www.forestry.gov.uk/pdf/WoodlandCarbonCode_V1.3_July2014.pdf/$FILE/WoodlandCarbonCode_V1.3_July2014.pdf. The involvement of government agencies can reduce transactions

intermediaries to conserve biodiversity on private land. An IUCN study in 2009[92] identified a range of issues in relation to contract design that must be considered by the parties entering a PES agreement. These include the 'project plan, identification of the parties, form and legal nature of the contract, objective and scope of the agreement, parties' rights and obligations, representations and warranties, payment structure, duration of the contract, risk allocation, remedies, and dispute settlement'.[93] Each of these issues would fall into one of two categories of the main problems associated with contract design when viewed from contract theory perspective: incentives and asymmetric information.[94] Furthermore, according to Stoneham and others, these problems are manifested as adverse selection, moral hazard and observability.[95] Other problems of contract design include commitment, credibility and incomplete contracts.[96] Commitment requires the parties to covenant in advance that they will stick to the activities agreed in the terms of the contract for a defined period.[97] The lack of such a pledge to keep to activities agreed in the terms of the contract reduces the credibility of the commitments made or their enforcement. Incomplete contracts reflect the fact that contracts typically take into account only the most relevant terms or those that can be easily verified by the courts in case of a breach. One of the reasons why most contracts are incomplete, as Salanié puts it, is because 'negotiating a contract is a costly business … [and] at some point the cost

---

costs, but can also weaken the voluntary element of the PES scheme which is seen as an attraction of this mechanism; Vijay K. Kolinjivadi and Terry Sunderland, 'A review of two payment schemes for watershed services from China and Vietnam: the interface of government control and PES theory' (2012) 17 *Ecology and Society* 10; Lasse Loft, Pham Thu Thuy and Cecilia Luttrell, *Lessons from Payments for Ecosystem Services for REDD+ Benefit-Sharing Mechanisms*, CIFOR Infobrief No. 68 (CIFOR, 2014).

92 Thomas Greiber (ed.), *Payments for Ecosystem Services: Legal and Institutional Frameworks* (IUCN, 2009).

93 *Ibid.* 45–62.

94 Contract theory is the study of how economic actors can and do construct contractual arrangements. For more on this, see Patrick Bolton and Mathias Dewatripont, *Contract Theory* (MIT Press, 2005).

95 Gary Stoneham and others, 'Auctions for conservation contracts: an empirical examination of Victoria's BushTender trial' (2003) 47 *Australian Journal of Agricultural Resources Economics* 477, 486.

96 Bernard Salanié, *The Economics of Contracts: A Primer* (2nd edn, MIT Press, 2005).

97 *Ibid.* 162.

of writing a specific clause to cover an unlikely contingency outweighs the benefit'.[98]

When designing PES contracts to conserve biodiversity, the non-fungible nature of biodiversity, where no two species or habitats are the same and therefore cannot be freely substituted for one another,[99] means that different types of contracts may be necessary if the functional value of biodiversity is to be achieved. These different contracts must be framed within a larger but local project/development plan which should normally be based on an overall national strategy. For biodiversity, this common goal would normally be set out in the countries' national biodiversity strategies, action plans and programmes, which are the principal instruments for implementing the Convention on Biological Diversity (CBD) at the national level.[100] The CBD requires countries that are parties to it to prepare a national biodiversity strategy and to ensure that this strategy is mainstreamed into the planning and activities of all those sectors whose activities can have an impact (positive and negative) on biodiversity.[101] There is considerable discretion as to how these instruments are designed but the overall goal must be in line with the objectives agreed at the international and regional levels.

While recognising the traditional conflict between development and conservation, especially in highly populated areas where the pressure to develop land that may be rich in biodiversity is highest, the challenge may be how to reconcile different PES contracts with the overall development plans and strategies. It may be the case that in order to meet the broader national goal, local biodiversity strategies have to be developed that take into account specific habitats and local circumstances. The national strategy may lay down the overall goal, while the need for different types of contracts suggests that it may also be necessary to develop an overall strategy of what the PES scheme aims to achieve. The PES contracts should, therefore, be designed to reinforce each other and work toward a common goal set locally, which will in turn contribute to meeting the national goal. For example, the Bolivian Los Negros-Santa

---

98 *Ibid.* 193–4.

99 E.g. one area of wetland is not the same as an area of sand dunes or an area of woodland. They support different ecological communities and contribute different ecosystem services. For more on exchangeability and equivalence of biodiversity, see Chapter 2, sections 2.3 and 2.4.

100 Convention on Biological Diversity, Art. 6. Working with such plans is also important for securing some coherence within the actions taken to conserve biodiversity; see Chapter 2, section 2.8.

101 *Ibid.*

Rosa PES scheme had as its overall goal to conserve biodiversity and protect the Los Negros river watershed. To achieve this, the buyers of the services (irrigators of Los Negros) negotiated two types of contracts with the providers of the services (the farmers of Santa Rosa). One type of the PES contract prohibited 'tree cutting, hunting and forest clearing on enrolled lands' and the other type focused on reforestation of deforested areas of land in the watershed.[102]

Closely linked to the necessity to develop an overall strategy for what the PES scheme aims to achieve is the importance of identifying all the parties – buyers, sellers or intermediaries – involved in PES transactions and the costs involved. Identifying all the parties is perhaps the first step in drafting individual contracts. A number of options are available, and the details of the contract will depend in large part on the nature of the parties. At the very least, in circumstances where the contracts have to be legally binding, the parties necessarily have to be legal persons who have the legal authority to enter into a contract. Deciding who has legal authority is very much linked to the issue of land tenure and property rights considered in section 3.4 above. Given the nature of biodiversity services and the need to adopt an ecosystem[103] and landscape[104] approach if the benefits derived from biodiversity are to be maximised, PES transactions will quite often involve a single buyer such as the government or public agency (acting within its wider remit to conserve biodiversity) or a private entity and multiple landowners or occupiers. In many countries, the state or some public bodies may have overall control over most biodiversity-rich private forests and lands. This suggests why the government or public agency may need to be a party to the PES contract, either as the party paying (usually through compensation schemes) or because of its rights which lie behind the more direct control exercised by occupiers.[105]

---

[102] Nina Robertson and Sven Wunder, *Fresh Tracks in the Forest: Assessing Incipient Payments for Environmental Services Initiatives in Bolivia* (CIFOR, 2005) 34.

[103] Convention on Biological Diversity (n. 4), COP 5 Decision V/6 (2000).

[104] Andrew F. Bennett, *Linkages in the Landscape: The Role of Corridors and Connectivity in Wildlife Conservation* (IUCN, 2003).

[105] E.g. in Cameroon, land that has not been registered as private or public property reverts to the common pool of 'national lands'. This land is administered by the state 'for the public good', and the law explicitly includes untitled lands occupied or used by rural communities by virtue of a customary tenure as 'national lands'; Nsoh, 'Customary Land Rights, Indigenous Rights and Land Expropriation in Cameroon' (n. 81) 283–4. For more examples of government

Furthermore, if we think of a PES scheme as a mechanism aimed at restraining land use practices that have a negative effect on the ecosystem, and not for incentivising activities to enhance services, then identifying which of the multiple landowners or occupiers to contract with, and the specific obligations for each, becomes even more significant. One challenge then is how to distinguish between those who provide high value services (which may well involve paying for improvements that help the services recover from past harmful activities) and those who provide low value services (which may well be those who refrain from land use practices that have a negative effect on the ecosystem), with a further consideration being the transactions cost of gathering the necessary information before any contract is offered.

How the scheme tackles these issues depends in large part on the intentions behind introducing the scheme, the motivation of the buyer(s) and the design and structure adopted. A number of design approaches can be considered.

3.5.1 General Subsidy

Schemes that adopt a flat subsidy approach (for example, the Environmental Services Payment (*Pago de Servicios Ambientales*, PSA) Program in Costa Rica),[106] cannot clearly distinguish between those parties who can provide high value services and those who provide low value services. Salzman has correctly noted that this is because they often operate by allowing any landowner or occupier within a particular area to participate regardless of whether they provide valuable services.[107] Indeed, the Costa Rican scheme was designed to compensate forest landowners for services provided by either planted or natural forest on their land. Of the four services recognised (greenhouse gas mitigation; hydrological services; scenic value; and biodiversity), the programme did not attempt to measure all four services on a given plot separately at once. Instead, a similar value for the bundled services was assumed to be provided by each hectare of enrolled land, regardless of the size of each

---

involvement in PES schemes, see Kolinjivadi and Sunderland, 'A review of two payment schemes for watershed services from China and Vietnam' (n. 91).

[106] For more on the Costa Rican programme, see FONAFIFO, *Environmental Services Payment (PSA) Program* (2015), available at www.fonafifo.go.cr/home/psa_eng/index.html; G. Arturo Sánchez-Azofeifa and others, 'Costa Rica's payment for environmental services program: intention, implementation, and impact' (2007) 21 *Conservation Biology* 1165.

[107] Salzman, 'Creating markets for ecosystem services' (n. 48) 918–19.

parcel[108] and whether the land could provide the required services to the allocated value. The policy operated on a 'first come, first served' basis; factors such as farm size, human capital and household economic level influenced participation in the programme, and large landowners were disproportionately represented among participants at the national and regional levels.[109] Allowing any landowner within a particular area to participate regardless of whether they provide valuable services may have the benefit of reduced costs in identifying which landowners or occupiers to enter into contracts with. Nevertheless, it is possible to have a general subsidy scheme that effectively encourages valuable service provision by having more precise eligibility requirements for participation,[110] but stricter requirements only add to the transaction cost of getting the information right.[111]

3.5.2 Direct Contracts

A mechanism that attempts to distinguish between those parties who can provide high value services and those who provide low value services is direct negotiation between the individual providers and the buyer. Direct negotiation is based on the assumption that different landowners can provide different levels of service and should therefore be compensated accordingly. The individual negotiations can take into account differing levels of service provision, and have the advantage of allowing individually crafted agreements. The Vittel PES programme in France is an example of a scheme in which the direct negotiation approach has been

---

[108] Sánchez-Azofeifa and others, 'Costa Rica's payment for environmental services program' (n. 106) 1166.

[109] *Ibid.* 1166; Simon Zbinden and David R. Lee, 'Paying for environmental services: an analysis of participation in Costa Rica's PSA program' (2005) 33 *World Development* 255.

[110] See e.g., the eligibility requirements for US Conservation Reserve Program (CRP) in Box 3.2 above and for EU agri-environment schemes which allow any land manager with land to apply for payments, but tenants must discuss their proposed application with the landlord to make sure it does not break the conditions of their tenancy; see text to n. 16 above; Environmental Stewardship (England) Regulations 2005, SI 2005/621; Rural Development Contracts (Land Managers Options) (Scotland) Regulations 2008, SSI 2008/159; Scottish Government, *Rural Development Contracts: Land Managers Options (LMO) Guidance 2008* (Scottish Government, March 2008) 1, available at www.scotland.gov.uk/Resource/Doc/217663/0058305.pdf.

[111] Salzman, 'Creating markets for ecosystem services' (n. 48) 918–19.

used.[112] The objective of the programme was to provide a high level of water quality by reducing nitrate rates in the aquifer. To achieve this objective, the programme has persuaded farmers 'to reconvert to extensive low-impact dairy farming, including abandoning agrochemicals, composting animal waste and reducing animal stocks in exchange for payments'.[113] This is achieved through directly negotiated long-term contracts (18–30 years) in which the payments vary according to the service provision cost on a farm-to-farm basis, and both land use and water quality are closely monitored over time. There are clearly benefits for recognising the different levels of service provider by each farm (for example, it creates incentives for farmers to use their private knowledge to optimise their level of action), but such an approach still requires identifying which landowners to negotiate with and can be labour intensive and costly if carried out with a large number of land managers. Moreover, because of the individual nature of the negotiations, it may also be hard to agree to PES contracts that reinforce each other and work toward a catchment-wide strategy for service provision measures that is framed within a larger but local project/development plan. Achieving the catchment-wide strategy, though, requires more landowners to be persuaded to join the scheme (or to sell their lands to someone willing to participate), which is only likely to add to the costs of operating the scheme.[114]

---

[112] Danièle Perrot-Maître, 'Private sector incentives for biodiversity conservation: the case of Vittel, France' in International Workshop on the Removal and Mitigation of Perverse, and the Promotion of Positive, Incentive Measures, Paris, October 2009, *Report*, UNEP/CBD/SBSTTA/14/INF/26 (6 April 2010) 41–3, available at www.cbd.int/doc/meetings/im/wsim-03/official/wsim-03-sbstta-14-inf-26-en.pdf; Danièle Perrot-Maître, *The Vittel Payments for Ecosystem Services: A "Perfect" PES Case?* (IIED, 2006).

[113] Irina Prokofieva, Sven Wunder and Enrico Vidale, *Payments for Environmental Services: The Way Forward for Mediterranean Forests?*, EEF Policy Brief No. 7 (2012) 1, 6; Danièle Perrot-Maître, 'The Vittel case: a public-private partnership in the mineral water industry' in FAO (ed.), *Case Studies on Remuneration of Positive Externalities (RPE)/Payments for Environmental Services (PES)*, prepared for the multi-stakeholder dialogue 12–13 September 2013), available at www.fao.org/fileadmin/user_upload/pes-project/docs/FAO_RPE-PES_Vittel-France.pdf.

[114] E.g. between 1988 and 2006, the number of farms in the Vittel programme declined from 37 to 26 while the average farm size increased to 150 hectares as the extensive production system required additional land and marginal farmers sold their land to the promoters of the scheme: *ibid.* 2.

3.5.3 Reverse Auction

Another approach that attempts to tackle the issue of the transaction costs of gathering the necessary information is the use of reverse auctions schemes. This approach relies on a publicised competition among landowners who provide sealed bids to the government or private entity of how much they are willing to accept for specified land use management changes.[115] Unlike the direct contract approach, the buyer is not concerned with identifying and distinguishing between which high value services and low value services providers to contract with, but rather with judging between the bids prepared and submitted by participants. Added benefits are the effective communication of goals to the target community, getting farmers to weigh the costs and benefits of land use changes, and leading landowners to change the ways they think about the benefits their land produces.[116] This approach recognises that the landowner is the party with the background of involvement with the site and who has control over access so as to enable or restrict detailed investigations. For large-scale features, the landowner may have the best knowledge, but they may not always be as knowledgeable about all the valuable features of their land, especially for the less obvious ones where only a few scientists might be able to identify the presence of plants or invertebrates which are ecologically significant but otherwise inconspicuous. The scheme therefore offers incentives for landowners to use their knowledge to optimise their level of action and join the scheme only if they can produce the biodiversity services targeted by the scheme.[117] But because the landowners may not hold all the information, transactions may be contemplated on the basis of a far from complete or equally shared understanding of the true position. In this regard, the reverse auction approach is perhaps best suited for bundled ecosystem services where there is only one buyer and multiple sellers.

Available evidence suggests that this approach can secure targeted action that retains and improves biodiversity benefits on private lands in a

---

[115] See e.g., Victoria State Government, *BushTender* (2014), available at www.depi.vic.gov.au/environment-and-wildlife/environmental-action/innovative-market-approaches/bushtender, and Brett Day and Laurence Couldrick, *Payment for Ecosystem Services Pilot Project: The Fowey River Improvement Auction*, Final Report prepared for the Department for Environment, Food and Rural Affairs (2013).

[116] Salzman, 'Creating markets for ecosystem services' (n. 48) 919.

[117] Gibbons and others, 'Should payments for biodiversity conservation be based on action or results?' (n. 64).

far more cost-effective manner than general subsidies because the successful bids are those that offer the best value for money.[118] For example, the BushTender programme of the State of Victoria, Australia, through which the government paid landowners to enter into three to six year contracts to adopt a range of vegetation management practices, has been hailed as the 'least cost' option for achieving biodiversity conservation benefits.[119] Nevertheless, the competitive nature of the bidding process in which landowners have to decide how much, or rather how little, they are willing to accept, increases the risk of economic arguments about services valued by humans in the short term outweighing non-economic justifications for ecosystem service provision in the short, medium and long term.[120] Beyond the significant environmental benefits registered,[121] wide acceptance of the schemes seems to be based on the fact that once land managers understand the non-market value of the services required by the schemes, they are much more willing to accept lower prices and internalise some of the perceived costs of conservation.[122] The problem with such calculations, though, is that as soon as another opportunity to provide greater returns emerges, land managers are less likely to continue to participate in the scheme, even those who might already have received payments, including for services that could take a long time to deliver

---

[118] Stoneham and others, 'Auctions for conservation contracts' (n. 95); Day and Couldrick, *Payment for Ecosystem Services Pilot Project* (n. 115).

[119] International Workshop Report (n. 112) 29–31. In fact, the apparent success of the Bush Tender led to the launch of the EcoTender programme in 2008, which is an auction-based approach that expands BushTender to include multiple environmental outcomes; Victoria State Government, *EcoTender* (2014), available at www.depi.vic.gov.au/environment-and-wildlife/environmental-action/innovative-market-approaches/ecotender. A similar approach has been replicated in many other schemes, see e.g., Alastair Grieve and Keith Uebel, *The NSW Environmental Services Scheme: Progress Report on Outcomes and Experience Developed during its Implementation* (NSW Government, 2003); Jim Binney and Charlie Zammit, 'Australia: The Tasmanian Forest Conservation Fund' in OECD, *Paying for Biodiversity: Enhancing the Cost-Effectiveness of Payments for Ecosystem Services (PES)* (OECD, 2010); Day and Couldrick, *Payment for Ecosystem Services Pilot Project* (n. 115).

[120] Kent H. Redford and William M. Adams, 'Payment for ecosystem services and the challenge of saving nature' (2009) 23 *Conservation Biology* 785.

[121] See e.g., Department of Sustainability and Environment, *BushTender: Rethinking Investment for Native Vegetation Outcomes – The Application of Auctions for Securing Private Land Management Agreements* (Department of Sustainability and Environment, 2008); Stoneham and others, 'Auctions for conservation contracts' (n. 95).

[122] Salzman, 'Creating markets for ecosystem services' (n. 48) 896–7.

(such as planting trees).[123] Even for the buyer who is not concerned with identifying and distinguishing between which high value services and low value services providers to contract with, but is only interested in the most cost effective bids, this approach raises the questions of when, and to whom, payments should be made for the provision of a service.

3.5.4 Paying a Third Party (Intermediary)

A final option for dealing with the issues of identifying the best sellers of service is to pay a third party, which could either be a local government, an existing non-governmental organisation or a specially-created funding body, rather than the landowners directly. The role of a third party is an important one and most successful PES schemes often involve them. Acting as 'honest brokers', third parties (intermediaries) can perform a variety of tasks,[124] such as helping sellers assess an ecosystem service 'product' and its value to prospective buyers; introducing buyers and sellers and building rapport between them; establishing ecosystem service baselines and the scope for additionality; identifying specific resource management interventions that will deliver service provision; aggregating multiple landowners/managers for more complex schemes; assisting in determining prices, accessing grants, structuring agreements and agreeing a mutually acceptable payment regime; activities related to implementation (including monitoring, certification, verification, etc.); and overall scheme administration.[125] In performing their role, the third party may become a party to a contract, not as a buyer or a seller, but instead as a 'neutral party' to build trust between the other parties. A 2006 report on the Vittel PES scheme in France concluded that '[t]rust-building through the creation of an intermediary institution (locally based and led by a "champion" sympathetic to the farmers' cause)' was a fundamental condition of success.[126] The involvement of a third party

---

[123] E.g. the New York City and Catskills Watershed scheme offered to pay farmers US$100 per acre every year for up to 15 years to plant native species along streams, but from January 1997 when the scheme started to the autumn of 2000, only about 400 acres had been planted; Daily and Ellison, *The New Economy of Nature* (n. 48) 80. Similarly in the Vittel PES scheme, the number of farms declined from 37 to 26 between 1988 and 2006; Perrot-Maître, 'The Vittel case' (n. 113) 2.

[124] Forest Trends, The Katoomba Group and UNEP, *Payments for Ecosystem Services, Getting Started: A Primer* (2008).

[125] Smith and others, *Payments for Ecosystem Services* (n. 32) 36.

[126] Danièle Perrot-Maître, *The Vittel Payments for Ecosystem Services: A 'Perfect' PES Case?* (IIED, 2006) 5.

who landowners can trust may be a reasonable strategy for guaranteeing that payments will be made to the landowners for their conservation activities once the PES contract takes effect, although it may be too administratively burdensome and add to the costs for smaller scale schemes.[127]

While the various approaches considered above attempt to deal with the issue of the cost of identifying the 'best' sellers of a service with varying degrees of success, the one issue that is yet to be resolved is the difficulty in determining how much to pay for particular actions and how this is factored into PES contracts. This remains one of the fundamental challenges in the design of PES contracts. As noted in section 3.3 above, most existing PES contracts are developed based on inputs rather than outcomes. It is probably easier and less costly to monitor compliance with an input-based contract (for example, payment for planting hedges), but inputs where costs are incurred do not guarantee service delivery (for example, the number of birds benefiting). In fact, it has been suggested that one of the reasons for the preference for input-based contracts is that there is no low-cost means of monitoring and measuring outcomes on which to construct ouput-based contracts.[128] The task of designing output-based contracts may be complex and expensive, but this should not discourage efforts to move away from the current predominant input-based approach to contract design to an output-based approach for at least part of the payments, if the risk of non-delivery is to be adequately attributed. The reported successes of the Tar-Pamlico project[129] and schemes in which some payments could vary based on the results of close monitoring throughout the term of the contract,[130] suggests that, at least for some services, output-based contracts can deliver at a reasonable cost.

---

[127] Salzman, 'Creating markets for ecosystem services' (n. 48) 920; Aske Skovmand Bosselmann and Jens Friis Lund, 'Do intermediary institutions promote inclusiveness in PES programs? The case of Costa Rica' (2013) 49 *Geoforum* 50.

[128] Stoneham and others, 'Auctions for conservation contracts' (n. 95) 487.

[129] EPA, 'North Carolina: Tar-Pamlico Basin' (n. 66).

[130] E.g. on the close monitoring of both land use water quality over time and how this ties in with payments in the Vittel PES scheme, see Prokofieva, Wunder and Vidale (n. 113) 6. See also Gibbons and others, 'Should payments for biodiversity conservation be based on action or results?' (n. 64).

3.6 GOVERNANCE

Good governance can mean different things to different organisations but would generally be concerned with how decisions are taken and implemented. The United Nations refers to good governance as promoting 'equity, participation, pluralism, transparency, accountability and the rule of law, in a manner that is effective, efficient and enduring'.[131] The European Commission recognises five principles that underpin good governance: openness, participation, accountability, effectiveness and coherence.[132] The OECD identifies a similar set of essential elements:

- openness, transparency and accountability;
- fairness and equity in dealings with citizens;
- efficient and effective services;
- clear and transparent laws and regulations;
- consistency and coherence in policy formation;
- respect for the rule of law; and
- high standards of ethical behaviour.[133]

According to Zaelke, Stilwell and Young, good governance is generally understood to be:

> more than a legal ideal and more than a development strategy. It also has been identified as a set of social norms comprising, among others, the rule of law, anti-corruption, and accountability. These good governance norms constrain the exercise of power in the public sphere by limiting the power of government, and in the private sphere by limiting market power and corporate control. These norms are concentrated in political institutions but also involve nongovernmental groups, including civil society, corporations and other businesses, and even capital markets. They are most effective when they are consistent with a community's general cultural values, the implicitly or explicitly shared abstract ideas about what is good and what is bad about society.[134]

---

[131] United Nations, *Global Issues: Governance* (2015), available at www.un.org/en/globalissues/governance/.

[132] European Commission, *European Governance: A White Paper* (Communication), COM(2001)428 final.

[133] OECD, *Final Report of the Ad Hoc Working Group on Participatory Development and Good Governance* (OECD, 1997) Pt 1, available at www.participatorymethods.org/sites/participatorymethods.org/files/final%20report%20on%20the%20ad%20hoc%20working%20gruop_OECD.pdf.

[134] Durwood Zaelke, Matthew Stilwell and Oran Young, 'Compliance, rule of law, and good governance' in Durwood Zaelke, Donald Kaniaru, and Eva

The different characterisations above suggest that good governance is therefore not only concerned with governmental activity but also private sector and non-governmental actors. Among the important elements of good governance are transparency, accountability and public participation[135] and these must be considered in the design and operation of any PES scheme. What is seen as most important, and how issues will be resolved, will vary greatly,[136] depending on the broader governmental and legal context, but at the minimum, there must be a means of ensuring that schemes are widely accepted as having the legitimacy to enable them to continue, and especially to justify any financial advantages granted to participants. Good governance helps to build trust and confidence among the various stakeholders, which is central to the success of any PES scheme.[137]

Kružíková (eds), *Making Law Work: Environmental Compliance and Sustainable Development* (Cameron May, 2005) vol. 1, 42.

[135] E.g. the Principles for the Advancement of Justice, Governance and Law for Environmental Sustainability, contained in the Rio+20 Declaration on Justice, Governance and Law for Environmental Sustainability agreed at the World Congress on Justice, Governance and Law for Environmental Sustainability, Rio de Janeiro, 17–20 June 2012, available at www.unep.org/rio20/Portals/24180/Rio20_Declaration_on_Justice_Gov_n_Law_4_Env_Sustainability.pdf.

[136] James Salzman and J.B. Ruhl, 'Currencies and commodification of environmental law' (2000–01) 53 *Stanford Law Review* 607, 668–93.

[137] Evidence suggests that the lack of trust can contribute to the failure of PES schemes. For example, in the Bolivian Los Negros-Santa Rosa PES schemes, the lack of trust between the sellers and buyers of the ecosystem services was a major obstacle to both sides agreeing to enter into formal contracts; Nigel Asquith and Maria Teresa Vargas, *Fair Deals for Watershed Services in Bolivia*, Natural Resource Issues No. 7 (IIED, 2007) 22; Robertson and Wunder, *Fresh Tracks in the Forest* (n. 102) 36. Similarly, the failure of the Brazilian *Proambiente* Programme has also been partly attributed to a lack of trust between the implementing parties and the participating parties; Paulo Sérgio Ferreira Neto, *Avaliação do Proambiente Programa de Desenvolvimento Socioambiental da Produção Familiar Rural* (Brasília, 2008) cited in Greiber, *Payments for Ecosystem Services* (n. 92) 91. A lack of trust has also been identified as a key barrier in a number of the DEFRA PES pilots; DEFRA, *Defra Payments for Ecosystem Services (PES) Pilot Projects* (n. 49) 30.

3.6.1 Transparency

A conservation policy based more strongly on privatised techniques such as those adopted under PES schemes might lack transparency.[138] If the main buyer of services is the government or a statutory body, then key information will be held at a single point and may well be available either through annual reports, accounts or other publications, or under freedom of information rules.[139] Even here, though, while legislation in some countries may provide for transparent decision-making and access to information, other countries (especially in the developing world) do not yet have the adequate statutory reporting rules that most developed countries have. Even for some countries that already have such rules, there is a discrepancy between legislative requirements and implementation in practice. As a result, environmental information, once collected by the government, tends to be difficult to obtain.[140] On the other hand, if the services are being bought by the private sector, then the transactions will not be subject to such disclosure rules and if multiple buyers are involved, the information will be scattered. This may suggest a need for further measures to be put in place to ensure the collection of data, transparent decision-making and access to the information, but these measures must be balanced against the needs of private entities to maintain some confidentiality of potentially market-sensitive information. The structure and effectiveness of arrangements for the collection of and

---

[138] Greater transparency has been identified as a key element in avoiding what has been described as an 'Ecosystem Service Curse': Jakub Kronenberg and Klaus Hubacek, 'Could payments for ecosystem services create an "ecosystem service curse"?' (2013) 18 *Ecology and Society* 10.

[139] E.g. the specific rules on access to environmental information under the Convention on Access to Information, Public Participation in Decision-Making and Access to Justice in Environmental Matters (Aarhus Convention), adopted on 25 June 1998, entered into force on 30 October 2001. See also Directive 2003/4/EC of the European Parliament and of the Council of 28 January 2003 on public access to environmental information and repealing Council Directive 90/313/EEC, [2003] OJ L41/26.

[140] E.g. art. 93 of Bolivia's Environmental Statute (Ley de Medio Ambiente No.1333 of 1992, *Gaceta Oficial de Bolivia* No. 1740) stipulates that environmental information held or used by the government at the national and departmental level must be public and accessible to any interested person. However, this is often not realised in practice and it can be very difficult to obtain information from the government once data has been collected; Asquith and Vargas, *Fair Deals for Watershed Services in Bolivia* (n. 137) 13 and 21.

access to information is something to bear in mind when PES schemes are being designed.

Where there is discrepancy between legislative requirements and implementation, transparency in PES schemes can be achieved in different ways. A recent study of PES experiences in Costa Rica over the last 20 years shows that transparency can be integrated at various stages throughout the lifecycle of the PES scheme, which must include clarity for the parties concerned, for example, by utilising non-differentiated payments that pay a fixed amount to every participant[141] and sharply defined over-the-counter transactions.[142] Wider transparency in PES schemes could also be achieved through public hearing processes such as workshops held throughout the lifecycle of the programme. For example, in the Bolivian Los Negros project, periodic workshops at local and national levels were held on a regular basis, and brought together project developers, local and national authorities, and other interested institutions.[143] The workshops were a forum for sharing experiences and lessons learned. Other ways in which transparency could be achieved is through the use of websites and making available the contact details of stakeholders for those interested in obtaining more information.

Another essential element of the proper functioning of any PES scheme is monitoring, reporting and verification requirements. Once a PES contract has been agreed, and the project is underway, it is not only important to manage it effectively, but it must also be consistently

---

[141] Ina Porras and others, *Learning from 20 years of Payments for Ecosystem Services in Costa Rica* (IIED, 2013) 17.

[142] The PES contracts are done through over-the-counter transactions, where the buyers purchase Certificates of Ecosystem Services, and choose the ecosystem service they wish to pay for, whether watershed services, biodiversity, landscape beauty or carbon sequestration. Such over-the-counter transactions are seen to offer transparency and convenience, by allowing the buyer to purchase their preferred ecosystem service and/or geographical area; *ibid.* 13–14.

[143] These are also some of the methods used in Brazilian Project Oasis and Extrema Water Steward Programme. It is also reported that the representatives of both the Oasis and Extrema Projects are quite receptive and willing to share information and experiences from these initiatives; Greiber, *Payments for Ecosystem Services* (n. 92) 65–6. The Brazilian government also recognises that only '[d]ialogue and cooperation among the private sector, civil society and governments will provide the answer for how to advance with PES in Brazil at scale': Kelli Barrett, 'Brazilian ecosystem services matrix brings transparency to environmental finance', *Ecosystem Marketplace*, 5 June 2015, available at www.ecosystemmarketplace.com/articles/brazilian-ecosystem-services-matrix-brings-transparency-to-environmental-finance/.

monitored and evaluated for service delivery and adequate distribution of benefits in accordance with the terms contained in the agreement. Obviously, there are challenges to how PES contracts can be monitored in a cost-effective manner, as noted in section 3.5 above. Nonetheless, irrespective of the contract design approach adopted (input- or output-based), making the verification results available to any buyers, intermediary institutions and the public will also increase transparency and legitimacy, as well as facilitating adaptive management processes which are central to the long-term success of the scheme.[144]

Further elements come into play where schemes are contemplated in particular situations. For example, the notion of free prior informed consent is now a key principle in international law related to indigenous peoples, and provides another important measure of transparency.[145] This requires non-coercive negotiations between companies or governments and indigenous peoples prior to development or other projects being undertaken. The requirement stipulates that a community has the right to give or withhold its consent to proposed projects that may affect the lands they customarily own, occupy or otherwise use.

3.6.2 Accountability

Closely linked to transparency is the requirement for accountability, which is again affected if more of the task of achieving conservation is passed from centralised bodies to the cumulative impact of many individual transactions. There are two elements to this: accountability for the overall goal of PES project and accountability between individual parties. On the one hand, under such a more fragmented system, it is harder for anyone to be held to account if the system fails to achieve its overall objectives. On the other hand, parties entering into PES contracts would need assurances that the terms of the contracts will be respected and that, in the event of non-compliance, their rights will be upheld.

To overcome this, some oversight body that fits within the wider institutional set-up that supports the PES programme might be put in place.[146] The legal basis of the involvement and roles of such a body must be clarified. Beyond issues which impact on institutional performance, such as the funding of such institutions, transaction costs of

---

[144] Forest Trends, The Katoomba Group and UNEP, *Payments for Ecosystem Services, Getting Started* (n. 125) 52–4.

[145] United Nations Declaration on the Rights of Indigenous Peoples, A/RES/61/295 (13 September 2007), art. 19.

[146] Greiber, *Payments for Ecosystem Services* (n. 92) 37–44.

running them, etc.,[147] establishing any structure that gives such a body a grip strong enough to be accountable for the overall performance of a scheme would at the same time diminish the flexibility and distributed decision-making that is a strength of moving to a more privatised approach. Ensuring accountability may mean that public authorities have to retain a strong role, albeit increasingly as regulator of others rather than as direct buyers of biodiversity services, but this may be complemented by private institutions which can bring more flexibility and independence, and important external capacities, as well as additional financial resources.[148] A functioning system for accountability would ensure that, even in government-funded programmes, private parties are protected from arbitrary decisions, and offer assurances that the terms of agreements will be applied and implemented in a fair manner, free from the influences of corruption and special interest groups.[149] Without the necessary safeguards in place to ensure accountability, parties will be less willing to enter into agreements.

3.6.3 Public Participation

Public involvement is another important element of good governance, and ensures that relevant stakeholders are involved, participating in decision-making in a meaningful way. The requirement for public involvement is grounded in a number of international instruments. For example, Principle 10 of the Rio Declaration provides that:

> Environmental issues are best handled with participation of all concerned citizens, at the relevant level. At the national level, each individual shall have

---

147 Salzman, 'Creating markets for ecosystem services' (n. 48) 882–3 and 920–26.

148 Smith and others, *Payments for Ecosystem Services* (n. 32); European Commission Directorate-General Environment, 'Payments for ecosystem services' (2012) 30 *Science for Environment Policy, Thematic Issue* 1, available at http://ec.europa.eu/environment/integration/research/newsalert/pdf/30si_en.pdf.

149 For a discussion on how accountability could be achieved in Mexico's Programme of Payments for Carbon, Biodiversity and Agro-forestry Services (PSA-CABSA), see Esteve Corberaa, Carmen González Soberanisc and Katrina Brown, 'Institutional dimensions of payments for ecosystem services: an analysis of Mexico's carbon forestry programme' (2009) 68 *Ecological Economics* 743, 759. For a more general discussion on the institutional and political theory issues, see Roldan Muradian and others, 'Reconciling theory and practice: an alternative conceptual framework for understanding payments for environmental services' (2010) 69 *Ecological Economics* 1202.

> … the opportunity to participate in decision-making processes. States shall facilitate and encourage public awareness and participation by making information widely available.[150]

There are two dimensions to public involvement: direct involvement and wider public participation. Each of these contributes to ideas such as shared responsibility[151] and stewardship (especially in relation to land)[152] which form part of the renewed relationship with the natural world required in the quest for a more sustainable future. PES schemes, like the other mechanisms under consideration in this book, by their very nature offer more by way of direct involvement to a wide range of participants than does the more traditional approach relying on public bodies dominating conservation activity. At present, decisions as to which species and habitats to protect and the form of protection to be given lie in the hands of public authorities who also provide the resources to support most of the conservation activity. Some PES schemes may force some beneficiaries of ecosystem services to pay for these, rather than relying on voluntary participation, but do at least connect them directly with the environment that affects them. Moreover, payments may offer an incentive to conserve the natural environment for features beyond the most threatened aspects of biodiversity which tend to be the focus of official conservation policy. In these ways, the ideals of shared responsibility and stewardship are further strengthened, as opposed to conservation being a matter that is exclusively in the hands of the public authorities. The involvement of more parties also offers opportunities, albeit also challenges, for decision-makers to gather important information about the needs and concerns of relevant stakeholders and may also introduce new and creative ideas to the programme design. Such information can lead to a greater integration between conservation and other policies, which is another desirable element of good environmental governance.[153]

Under existing approaches to nature conservation, a fundamental feature remains that conservation decisions are mostly made by the authorities themselves, with public participation being an incidental consequence of the interaction with other decision-making processes

---

150 Rio Declaration on Environment and Development, June 1992, Principle 10, UN Doc. A/CONF.151/26, vol. I, (1992) 31 ILM 874. See also Aarhus Convention, Arts 1 and 2.

151 Declaration of the United Nations Conference on the Human Environment, Stockholm 1972, UN Doc. A/Conf.48/14/Rev 1 (1973), Preamble, para. 7.

152 See n. 71 above.

153 UNEP, *Environmental Governance* (UNEP, 2015), available at http://unep.org/pdf/brochures/EnvironmentalGovernance.pdf.

rather than an integral part of conservation regulation. However, experience suggests that when it comes to the participation of the public as stakeholders, they are more likely to support decisions in which they feel vested.[154] There is a danger, though, depending on the contract approach adopted for any particular scheme, that the role of the wider public as stakeholders might be even more marginalised. PES schemes that adopt direct negotiations in order to take into account differing levels of service provision, would be a matter of private negotiation between the buyer and each individual party to the agreement. Even though they would have to fit within the overall objective of the PES scheme, money changing hands for ecosystem services would bring arrangements within the commercial and financial spheres where public participation is generally more limited at the individual level. By moving away from public law mechanisms the opportunities for standard approaches to public participation are reduced.[155] Such participation is linked to the direct action or regulatory decisions of a public authority and will diminish if conservation effort relies less on such intervention and more on the aggregation of many individual decisions in the private sector and of individual transactions between private parties themselves or between private parties and the regulatory bodies.[156] A survey of landowners, including both participants and non-participants in the Costa Rican PES scheme, concluded that 'farm size, human capital and household economic factors, and information variables significantly influence participation' in the programme.[157] This shows that the importance of public participation is most evident in PES during initial scoping and negotiations, which may help in identifying potential buyers and sellers through publicising the programme, as well as serve to convince a broader range of stakeholders

---

154 See e.g., Nicolas Kosoya, Esteve Corberab and Kate Brown, 'Participation in payments for ecosystem services: case studies from the Lacandon rainforest, Mexico' (2008) 39 *Geoforum* 2073.

155 On the role and value of public participation see Frans H.J.M. Coenen (ed.), *Public Participation and Better Environmental Decisions: The Promise and Limits of Participatory Processes for the Quality of Environmentally Related Decision-making* (Springer, 2009).

156 An evaluation of a PES scheme in Mexico has shown that 'the nature of PES rules and the effectiveness of communication with government officers and NGOs influence resource managers' ability and willingness to participate': Kosoya, Corberab and Brown, 'Participation in payments for ecosystem services' (n. 154) 2073; Christian Schleyer and others, 'Opportunities and challenges for mainstreaming the ecosystem services concept in the multi-level policy-making within the EU' (2015) 16 *Ecosystem Services* 174.

157 Zbinden and Lee, 'Paying for environmental services' (n. 109) 255.

of the benefits that PES transactions can offer. At the scoping stage, the designers must consider how the public will be engaged (via closed meetings or open public meetings), which members of public to engage (individual stakeholders, broad groups, or special groups created specifically for the PES scheme), and for how long the public will be engaged (a single meeting or a series of meetings held over a long period of time).[158] This process must therefore go hand-in-hand with the process of identifying buyers, sellers and any intermediaries.

3.7 ETHICS: ROLE OF THE MARKETS AND PUBLIC GOODS

At the centre of PES (as with the other mechanisms considered in this book) is an approach that applies market-based ideas to our dealings with nature, often involving human intervention to create or enhance habitat in return for payments. Besides the technical challenges involved in designing an effective PES scheme and ensuring its successful implementation, there is also the more fundamental ethical issue that questions the legitimacy of this approach. The public goods character of biodiversity means that the choice of regulatory mechanisms is always going to be hotly contested and opting for a market-based approach involves accepting certain value-based views about the human relationship with nature

---

[158] In the Bolivian Los Negros-Santa Rosa PES pilot project, Fundación NATURA (a non-governmental organisation) conducted a series of negotiations and public campaigns to engage potential ecosystem services buyers and sellers; Asquith and Vargas, *Fair Deals for Watershed Services in Bolivia* (n. 137); Robertson and Wunder, *Fresh Tracks in the Forest* (n. 102). Similarly, in the Brazilian Proambiente programme it took several years of public discussion, commencing with rural social movements in the Amazon and in time advancing to the level of a federal policy; Paulo Sérgio Ferreira Neto, *Avaliação do Proambiente Programa de Desenvolvimento Socioambiental da Produção Familiar Rural* (Brasília, 2008) cited in Greiber, *Payments for Ecosystem Services* (n. 92) 91. In the Fowey River Improvement Auction pilot project, the contractors identified organisations that were either directly connected to the Fowey (local groups), closely connected to the Fowey (area groups) or loosely connected to the Fowey (regional groups) and approached them to agree to a telephone interview. The interviews also included a wide variety of stakeholders from different entities in the catchment, including town councils, anglers, boating interests, Fowey Harbour Commission, users of the port; Day and Couldrick, *Payment for Ecosystem Services Pilot Project* (n. 115) 63.

that only adds to difficulties.[159] In most PES schemes, the widespread use of input-based valuation methods is, perhaps, acknowledgement that elements of biodiversity are not interchangeable and their value cannot be captured wholly in monetary terms.[160] Yet to some, the very notion that elements of biodiversity can be treated as having a monetary value, or even be subject to the standard rules of private property, are abhorrent from some perspectives.[161] These ethical views and more, including the wider concern over the distorting effects and limits of markets issues, are considered more fully in Chapter 8.

Beyond the view that biodiversity should not be regarded as a commodity, an underlying difficulty with using PES schemes to promote biodiversity conservation is the question of which landowners should be supported by ecosystem service payments: those who currently provide services or those whose properties pose the greatest problems (and hence the greatest potential for increased service provision) or every landowner who provides ecosystem services – and how much? Answering these questions is at the very heart of what PES stands for. If there is a valuable ecosystem service to be obtained from the land, should landowners whose actions may diminish these be seen as polluters, and therefore subject to the polluter pays principle, which underpins modern environmental law and policy, and requires the polluter to bear the cost of preventing, controlling and cleaning up harm to the environment?[162] If a farming activity is causing harm to biodiversity, then the farmer might be seen as a polluter and hence, as Salzman puts it, 'they presumably should not be paid, but regulated or taxed'.[163]

However, such characterisation will depend on the view taken of how far the legal limits on what landowners can and cannot do on their lands are to be seen as setting the limits of who is to be judged to be a 'polluter'. There is a difference between what is deemed to be unacceptable, and hence unlawful (or at least subject to punitive taxation), and

[159] Jerneja Penca, 'Marketing the market: the ideology of market mechanisms for biodiversity conservation' (2013) 2 *Transnational Environmental Law* 235.

[160] Robert Constanza and others, 'Changes in the global value of ecosystem services' (2014) 26 *Global Environmental Change* 152, 157.

[161] Cormac Cullinan, *Wild Law: A Manifesto for Earth Justice* (2nd edn, Green Books, 2011) 147–8.

[162] Margaret Rosso Grossman, 'Agriculture and the polluter pays principle: an introduction' (2006) 59 *Oklahoma Law Review* 1, 1–39; Ved P. Nanda, 'Agriculture and the polluter pays principle' (2006) 54 *American Journal of Comparative Law* 317.

[163] Salzman, 'Creating markets for ecosystem services' (n. 48) 934.

what is undesirable, but not prohibited; after all, virtually all human activity has an adverse effect on some aspect of biodiversity. This leaves a gap where PES schemes can be proposed as a mechanism to influence landowners away from the publicly undesirable (but not unlawful) choices that they make on their land and towards the more desirable ones. Indeed, a further role for PES may exist where conduct is not appreciably undesirable overall, but is harmful to a specific interest, such as where a water company might want to restrict within their catchment activities which are polluting by their very high standards but not generally a cause for any concern. Rather than dealing with the consequences afterwards, it may be appropriate (and ultimately cheaper) to allow the water company to pay the landowners to curb the activities which are problematic in the specific circumstances (but neither unlawful nor inherently problematic in the absence of special interests).[164]

The implications of this argument for payments in a biodiversity context are more complex, where the impact on the functional value of, for example, a wildlife corridor depends critically on adjoining parcels of land. For a start, this demonstrates how the traditional property-based legal treatment of land and nature does not match the ecosystem approach that is now regarded as the basis for effective conservation. In calling for 'adaptive management to deal with the complex and dynamic nature of ecosystems',[165] the ecosystem approach also places an emphasis on ecosystems at all scales: 'a grain of soil, a pond, a forest, a biome or the entire biosphere'.[166] Property law, however, fragments the environment into separate parcels which are treated independently, ignoring their place within an integrated ecosystem. Natural features do not determine the boundaries of most units of land, and, except where there is interference with the legal rights held by others, each unit can be managed with no regard for the world beyond its boundaries.[167] The

[164] An analogy might be that property law protects a minimum 'right to light' but if a property owner wants to secure more than this, or a view, that must be acquired through agreements with the relevant neighbours, who are entitled to seek payment for this additional restriction on how they use their land.

[165] Convention on Biological Diversity (n. 4), COP 5 Decision V/6 (2000), para. 3.

[166] *Ibid.*

[167] The need to overcome this underpins the emerging concept of connectivity conservation; Barbara Lausche and others, *The Legal Aspects of Connectivity Conservation: A Concept Paper* (IUCN, 2013). This is also an area where greater legal recognition of ecosystem services might benefit conservation, recognising activities diminishing ecosystem services as a form of damage to the property that would benefit.

extension of property-based concepts to other elements of biodiversity, it might be argued, will only exacerbate the problems of holdouts and free-riders.[168]

The functional value of biodiversity habitat depends critically on adjoining parcels of land.[169] If successful, the benefits from the sum of land parcels managed for biodiversity conservation should be greater than its parts. This would be in line with the ecosystem approach, which PES is an aspect of. This can be frustrated, though, by the actions of some who are holding out for a higher than market price or just not interested in taking part in a voluntary scheme that requires changes to how they manage their lands. Without their participation, it may be impossible to create effective habitats. Moreover, neighbours of those who dedicate their lands to biodiversity conservation may choose not to conserve biodiversity on their own land but, instead, free-ride on the wildlife amenities on adjacent land. The problems of holdouts and free-riders are most easily seen in the context of biodiversity conservation but likely much less of a concern in the context of other services, such as water purification, because the effectiveness of landscape management, for example, is less likely to turn on the actions of a handful of landowners.

For a public good such as nature, preventing those who have not contributed to the enhancement of the service from benefiting or finding ways to get them to pay will be a huge challenge to the designers of any PES scheme. For the schemes based on a general subsidy model such as the EU agri-environment schemes and the Costa Rican *Pago de Servicios Ambientales* (PSA), paying every landowner who provides ecosystem services the standard market price calculated according to a fixed formula could be the best way to ensure that all neighbouring landowners needed for the effective delivery of the service are on board.[170] Such an approach, however, may limit the freedoms of individual landowners and

---

[168] For more on the problems of holdouts and free-riders in PES schemes, see Kronenberg and Hubacek, 'Could payments for ecosystem services create an "ecosystem service curse"?' (n. 138); Andrea Baranzini, Anne-Kathrin Faust and David Huberman, 'Household demand for international ecosystem services: a Swiss case study' in Joshua Bishop and Chloe Hill (eds), *Global Biodiversity Finance: The Case for International Payments for Ecosystem Services* (Edward Elgar Publishing, 2014) 57; W.K. Jaeger, *Ecosystem Services and the Potential Role for Markets*, Extension Service Paper No. EM 9033 (Oregon State University, 2011) 4.

[169] See Chapter 2, section 2.8.

[170] Although there are still problems of compliance and enforcement to contend with, it could be argued that accepting a certain degree of wasted payment is a price worth paying for a simple scheme.

the flexibility that is an attraction of the scheme. For schemes that are based on direct individually negotiated agreements or reverse auctions, there is a danger that some landowners who are holding out for higher prices than the market price may not participate, and this would have a detrimental effect on the functional value of the biodiversity service. As noted throughout this book, though, the benefits of any approach should not be overlooked simply because it is not perfect.

3.8 CONCLUSION

The basic idea of using PES schemes to enhance biodiversity conservation is quite simple. However, establishing a legally binding framework to underpin it into the future and take account of the likely interests of the various parties is not straightforward. There is a substantial volume of literature highlighting the challenges and suggesting ways for overcoming them. Despite the challenges involved, the use of PES schemes has continued to expand in many places around the world,[171] and there is evidence that where they have been adopted to promote biodiversity conservation, they are delivering real benefits to biodiversity in a cost-effective manner.[172] However, if they are to form part of our wider biodiversity conservation policy that would speed up the achievement of internationally agreed goals on biodiversity conservation,[173] then closer attention must be paid to the challenges they present.

The nature of biodiversity requires long-term commitment in most cases to deliver any real benefits. As is the case with the other mechanisms considered in this book, such long-term commitments make it all the more important to have legal guarantees in place, but also more problematic to establish them. Over time, the priorities, capacities, desires and even identities of the key parties are liable to change and recollections of what was agreed and any compromises it embodies will dim. Such considerations point towards the benefits of having in place clear and enforceable obligations in order to build confidence that the biodiversity benefits paid for are going to be created and maintained at least throughout the duration of the agreement and ideally beyond. Yet focusing on just how such rights, responsibilities and obligations can be

---

171 Bennett, Carroll and Hamilton, *Charting New Waters* (n. 13).

172 Stoneham and others, 'Auctions for conservation contracts' (n. 95); Perrot-Maître, *The Vittel Payments for Ecosystem Services: A "Perfect" PES Case?* (n. 112).

173 UNEP, *Global Environment Outlook (GEO) 5* (n. 6).

structured and guaranteed within an agreement can be an obstacle to any progress, given the need to make the scheme attractive and to keep landowners motivated enough to continue to participate so that the biodiversity benefits can be delivered.

Further elements in common with the other mechanisms considered in this book relate to valuation and eligibility for payments. By calculating payments based on inputs and on land area, rather than on outcomes (or the provision of the service in the medium to long term), many of the issues of eligibility are addressed in existing schemes. However, this is not necessarily fair to the successors in title who may be responsible for maintaining services even though payments had already been made, or to those already providing services voluntarily, without the need for further expensive management activity to enable their land to deliver these benefits. Whatever approach is adopted, an initial question is to clarify the objective being sought, in line with the ecosystem approach.[174] When paying for ecosystem services, it is important to define exactly what service is being paid for, especially as these may be difficult to disaggregate, for example, valuable habitat provided to further biodiversity may also act as a carbon sink.

Further problems arise from the ways in which so much of the land is held and used around the world. There is a lack of clarity over how managing land to deliver this new 'produce' in the form of ecosystem services fits within existing structures of landholding based on agricultural and forestry production. To what extent can using land for the provision of ecosystem services fall within definitions of agriculture and good husbandry?[175] Who is the person entitled to sell ecosystem services? In the use of PES schemes, tenancy arrangements pose a challenge, both in ensuring that any benefits are delivered in the long term and in matching the payments made to the actual contributions of various parties. Complications also arise in areas with multiple occupancies where no one individual can take sole control or dispose of the resources, yet the interests and actions of each individual are immensely powerful – potentially beneficial or destructive. Similar issues arise where difficulties on how to reconcile traditional customary land tenure systems create uncertainty as to who owns land, yet the scheme must carefully balance

---

[174] Thomas Greiber and Simone Schiele, *Governance of Ecosystem Services: Lessons Learned from Cameroon, China, Costa Rica, and Ecuador*, IUCN Environmental Policy and Law Paper No. 79 (IUCN, 2011) 131–2.

[175] This may be important, e.g. in relation to tax liability (see Chapter 6) and in determining whether a tenancy agreement falls within the distinct rules for agricultural tenancies (Agricultural Holdings (Scotland) Acts 1991 and 2003).

the rights/responsibilities and obligations of the community of users if the biodiversity benefits from the land/forest are to be sustained.

Given the public good character of biodiversity, there are particular difficulties in how to balance the public interest in nature conservation (sought by the use of PES) with the private interests (property rights) of landowners, tenants and others across the range of different land-holding patterns. For this reason, the greatest potential for using PES schemes to deliver biodiversity benefits may be realised when they sit alongside other existing types of funding and regulation, including minimum standards of stewardship.[176] Wider uptake must also take into account the local circumstances, and will require, at the very least, a well-defined payment system, flexible enough to accommodate different layers of ownership and use and to cope with the changes in these over time. This will be necessary in order to secure the provision of biodiversity services over the decades that are needed if substantial public benefits are to be achieved. In this regard, and moving forward in addressing many of the risks and difficulties with PES schemes, there is scope to try out various approaches and perhaps place-based approaches, that is, ones that reflect the principles of an ecosystems approach in a local area.

[176] It has also been pointed out that while the conservation goals may be laudable, the trade-distorting nature of some general subsidies also needs to be considered; UNEP, *Global Environment Outlook (GEO) 4: Summary for Policy Makers* (UNEP, 2007) 29.

4. Biodiversity offsetting

The basic idea behind biodiversity offsetting is that activities harming biodiversity in one place can be accepted so long as they are balanced by activities enhancing biodiversity elsewhere. The aim is to ensure that overall there is no net loss to biodiversity. In this way, it is argued, the needs of both conservation and development can be reconciled, enabling the economic development required by society to proceed whilst ensuring that biodiversity does not suffer.

> Biodiversity offsets are measurable conservation outcomes resulting from actions designed to compensate for significant residual adverse biodiversity impacts arising from project development after appropriate prevention and mitigation measures have been taken. The goal of biodiversity offsets is to achieve no net loss and preferably a net gain of biodiversity on the ground with respect to species composition, habitat structure, ecosystem function and people's use and cultural values associated with biodiversity.[1]

Such offsetting is already a feature of a number of legal and policy regimes (although not necessarily under that name or expressly recognised as such) and is being more widely explored as a means of resolving conflicts between environmental and economic goals.[2] In the face of development that will cause harm to species or habitats, offsetting undoubtedly can make a contribution to ensuring that something is done to make up for the losses to biodiversity which would otherwise simply be ignored, but there is more debate over how effective this mechanism can be and how far it should be relied on.

Since offsetting is essentially a form of trading one thing for another (within limits), it is central to any market-based approach to environmental or other challenges. In the environmental context, offsetting has

---

[1] Business and Biodiversity Offsets Programme (BBOP), *Biodiversity Offset Design Handbook* (2009) 6.

[2] Joseph W. Bull and others, 'Biodiversity offsets in theory and practice' (2013) 47 *Oryx* 369; Bárbara Gonçalves and others, 'Biodiversity offsets: from current challenges to harmonized metrics, review article' (2015) 14 *Current Opinion in Environmental Sustainability* 61.

achieved most prominence in relation to greenhouse gases, where national and international frameworks accept that an obligation on one party to reduce emissions may be satisfied by that party securing that equivalent reductions are achieved by another party (in some circumstances, at least).[3] Such an approach allows the total level of emissions to be reduced in a way that is more efficient and cost-effective and less disruptive than requiring each party individually to achieve the reductions within their own activities. Even here, though, there are challenges in providing mechanisms for guaranteeing the delivery of the offset and avoiding double-counting, whilst the non-fungible and often irreplaceable nature of biodiversity, and the long time-scales within which many natural processes operate, add further complications and possibly restrictions on when offsetting can work.

A number of biodiversity offsetting schemes are now in operation, or proposed, around the world, but the underlying idea also features in a number of other regimes. The best known example is probably the US wetland mitigation scheme, requiring compensatory measures to be taken where permits are given for development affecting certain wetland environments,[4] whilst schemes in Australia are also significant.[5] Recent proposals in England have given rise to detailed debate of the issues

---

[3] E.g., the Clean Development Mechanism under the Kyoto Protocol and the EU Emissions Trading Scheme. These remain controversial: Gerd Winter, 'The climate is no commodity: taking stock of the Emissions Trading System' (2010) 22 *Journal of Environmental Law* 1.

[4] For links to factsheets, rules and guidance from the US Environmental Protection Agency, see EPA, *Compensatory Mitigation* (2015), available at http://water.epa.gov/lawsregs/guidance/wetlands/wetlandsmitigation_index.cfm; a useful summary of the scheme and its development is in Philip Womble and Martin Doyle, 'The geography of trading ecosystem services: a case study of wetland and stream compensatory mitigation markets' (2012) 36 *Harvard Environmental Law Review* 229, 245–52.

[5] For the Queensland scheme see Queensland Government, *Environmental Offsets* (2015), available at www.qld.gov.au/environment/pollution/management/offsets/; and for New South Wales, see NSW Government, *Assessing and Offsetting Impacts on Biodiversity* (2015), available at www.environment.nsw.gov.au/biodivoffsets/index.htm; Anthea Coggan and others, 'Transferable mitigation of environmental impacts of development: two cases of offsets in Australia' (2013) 15 *Journal of Environmental Policy and Planning* 303.

involved,[6] whilst France[7] and Germany,[8] among others,[9] also have some form of offsetting schemes in operation. What are in essence offsetting arrangements can also be included as permit conditions or the like under various regulatory schemes where the requirements specified may fulfil a compensatory role.[10]

Beyond such schemes, two examples from EU law can be given, where although the specific language of biodiversity offsets does not appear the basic approach does. The first is under the Habitats and Species Directive,[11] where activities that damage the integrity of a site designated as part of the Natura 2000 network can be permitted only if certain demanding criteria are met[12] and 'all compensatory measures necessary

---

6 DEFRA, *Biodiversity Offsetting in England: Green Paper* (DEFRA, 2013) and the related papers and reports noted later in this chapter.

7 Fabien Quétier, Baptiste Regnery and Harold Levrel, 'No net loss of biodiversity or paper offsets? A critical review of the French no net loss policy' (2014) 38 *Environmental Science and Policy* 120.

8 Juliane Albrecht, Jochen Schumacher and Wolfgang Wende, 'The German impact-mitigation regulation: a model for the EU's no-net-loss strategy and biodiversity offsets?' (2014) 44(3) *Environmental Law and Policy* 317.

9 Becca Madsen, Nathaniel Carroll and Kelly Moore Brands, *State of Biodiversity Markets Report: Offset and Compensation Programs Worldwide* (Ecosystem Marketplace, 2010), available at www.ecosystemmarketplace.com/documents/acrobat/sbdmr.pdf; and Becca Madsen and others, *2011 Update: State of Biodiversity Markets* (Forest Trends, 2011), available at www.ecosystemmarketplace.com/reports/2011_update_sbdm; Institute for European Environmental Policy, *Policy Options for an EU No Net Loss Initiative: Final Report, Annexes*, ENV.B2/SER/2012/0028; ICF and GHK, *Exploring Potential Demand For and Supply of Habitat Banking in the EU and Appropriate Design Elements for a Habitat Banking Scheme* (Final report submitted to Directorate-General Environment, 2013). Further consideration of various examples, especially of the problems created by the need to find long-term arrangements, is provided in the Final Report by ICF Consulting Services for the European Commission: ICF Consulting, *Study on Specific Design Elements of Biodiversity Offsets: Biodiversity Metrics and Mechanisms for Securing Long Term Conservation Benefits, Final Report*, ENV.B.2/ETU/2013/0060r (Directorate-General Environment, 2014).

10 E.g. planning obligations agreed when planning permission is granted in the United Kingdom; Neil Collar, *Planning Law* (3rd edn, W. Green, 2010) 199–219.

11 Council Directive 92/43/EEC of 21 May 1992 on the conservation of natural habitats and of wild fauna and flora, [1992] OJ L206/7.

12 There must be 'imperative reasons of overriding public importance' and no alternative solution; *ibid.* art. 6(4).

to ensure that the overall coherence of Natura 2000 is protected'.[13] This is in effect requiring an offset to ensure no net loss to the coherence of the Natura network and may well require setting aside new sites for the creation or protection of equivalent habitat, although there are concerns over how thoroughly this requirement is carried through and policed in practice.[14] The second is the Environmental Liability Directive,[15] where in the case of 'biodiversity' damage,[16] the obligation on the party responsible is to return the site affected to its baseline condition, or if this is not possible to provide 'complementary remediation', the purpose of which is to 'provide a similar level of natural resources and/or services, including, as appropriate, at an alternative site' as would be provided by the original site in its baseline condition.[17] Again, what is essentially an offset arrangement is required, albeit as a second best option.

The fact that in both these examples the role of offsetting is secondary, coming into play only if the primary measures of conservation or restoration on site cannot be implemented, is important. This is a widespread approach which places priority first on preventing and then reducing or mitigating[18] as far as possible any harm to biodiversity before allowing offsetting to take effect: 'The role of biodiversity offsets is effectively as a "last resort", after all reasonable measures have been taken first to avoid and minimise the impact of a development project and then to restore biodiversity on-site'.[19] The greater the weight that is

---

13 *Ibid.* art. 6(4).

14 D. McGillivray, 'Compensating biodiversity loss: the EU Commission's approach to compensation under Article 6 of the Habitats Directive' (2012) 24 *Journal of Environmental Law* 417; Baptiste Regnery, Denis Couvert and Christian Kerbiriou, 'Offsets and conservation of the species of the EU Habitats and Birds Directives' (2013) 27 *Conservation Biology* 1335.

15 Directive 2004/35/CE of the European Parliament and of the Council of 21 April 2004 on environmental liability with regard to the prevention and remedying of environmental damage, [2004] OJ L143/56.

16 Defined in terms of harm affecting the favourable conservation status of protected species or habitats; *ibid.* art. 2.

17 *Ibid.* art. 6 and Annex II.

18 It is an unfortunate terminological quirk that one of the largest and most well-established offsetting schemes, compensating losses in one place with gains elsewhere, is the wetland mitigation scheme in the United States, whereas more generally 'mitigation' is used to refer to ways of reducing, not compensating for, the impact of development.

19 BBOP, *Biodiversity Offset Design Handbook* (n. 3) 12. See also the hierarchy in the *National Planning Policy Framework for England* (Department for Communities and Local Government, 2012) para. 118: 'if significant harm resulting from a development cannot be avoided (through locating on an

placed on offsetting as a means of achieving effective conservation, the more exact, robust and secure the mechanism must be. On the other hand, if it is deployed as a supplement to other, more traditional conservation measures, a more relaxed attitude can be taken, although there is always the danger that if offsetting is recognised at all as a means of departing from stricter direct conservation obligations (as is the case with the EU Habitats Directive), then this exception can be too widely used so as to weaken the primary requirements.

The role to be played by offsetting and its relationship with other conservation measures are thus vital. With many models around the world, offsetting can hardly be called an innovation, but what is new is the widespread interest in giving this mechanism a more significant role in the conservation enterprise. A strong market-based approach might call for the greater use of offsetting in place of restrictive conservation measures devoted to particular sites or features, but others would challenge the view underlying that approach, namely, that elements of biodiversity are widely exchangeable.[20] At the other extreme, denying the acceptability of any measures based on equivalence or exchangeability will lead to a total rejection of offsetting. In between, one might accept offsetting as a means of achieving at least some compensatory benefits to balance the harmful consequences for biodiversity in the case of projects approved on the basis of other benefits legitimately overriding the losses to nature: 'Biodiversity offsetting should only be utilized where development is appropriate, that is, it ought not to justify or excuse development projects that should not take place in the first place.'[21] This matches the

---

alternative site with less harmful impacts), adequately mitigated, or, as a last resort, compensated for, then planning permission should be refused'. The wetland mitigation scheme in the United States is similarly based on the hierarchy of 'avoid, minimize, compensate'; EPA, *Compensatory Mitigation Factsheet* (2015), available at www2.epa.gov/sites/production/files/2015-08/documents/compensatory_mitigation_factsheet.pdf. The EU's Working Group on No Net Loss of Ecosystems and their Services also firmly takes this view: European Commission, *Development of Operational Principles of Any Proposed EU No Net Loss Initiative, Final Version* (2013), available at http://ec.europa.eu/environment/nature/biodiversity/nnl/pdf/NNL_Operational_Principles.pdf.

[20] An analysis of consultation responses on biodiversity offsetting proposals in England has revealed fundamentally different value frameworks underlying the specific differing views expressed; S. Sullivan and M. Hannis, 'Nets and frames, losses and gains: value struggles in engagements with biodiversity offsetting policy in England' (2015) 15 *Ecosystem Services* 162.

[21] Peter Williams, 'Growing houses and trees: integrating biodiversity conservation and urbanisation – an Australian case study' (2014) 6 *International*

hierarchy captured[22] in the English National Planning Policy Framework which states that in terms of the natural environment 'if significant harm resulting from a development cannot be avoided (through locating on an alternative site with less harmful impacts), adequately mitigated, or, as a last resort, compensated for, then planning permission should be refused'.[23]

Across this range of views,[24] there is a large area where people may be willing to accept that in this imperfect world, where biodiversity is so often simply ignored, there can be a role for offsetting in achieving improvements on the present position, provided that the details of any scheme are right. People will disagree on when the details are right – and indeed, on whether means can be found for offsetting ever to deliver effective conservation.[25] The rest of this chapter is largely devoted to exploring the elements involved in designing offsetting schemes to satisfy the needs and concerns of various parties. The list of elements involved and the range of solutions can be arranged in many ways,[26] and this chapter tries to capture the key issues.

4.1 WHAT IS A BIODIVERSITY GAIN?

Before reaching the issue of how to measure equivalence in any situation and whether no net loss to biodiversity has been achieved,[27] there are preliminary questions of what constitutes a net loss and what can qualify as a gain for biodiversity which can be put into the balance against a loss. The latter issue of what counts as a 'loss' is itself not free from difficulty since any alteration in habitat or conditions will produce detrimental and

Journal of Law in the Built Environment 152, 160. Many offsetting schemes exclude particularly sensitive sites from their scope, as discussed below.

22 If not always thought to be rigorously applied.

23 Department for Communities and Local Government, *National Planning Policy Framework* (2012) para. 118.

24 Critical analyses of offsetting arrangements can come from many perspectives and disciplines, such as Helen Tregidga, 'Biodiversity offsetting: problematisation of an emerging governance regime' (2013) 26 *Accounting, Auditing and Accountability Journal* 806.

25 Susan Walker and others, 'Why bartering biodiversity fails' (2009) 2 *Conservation Letters* 149.

26 IUCN, *Biodiversity Offsets Technical Study Paper* (IUCN, 2014). See section 4.8 below.

27 Toby Gardner and others, 'Biodiversity offsets and the challenge of achieving no net loss' (2103) 27 *Conservation Biology* 1254.

beneficial effects for different species. A change in the environment can be seen as bringing a gain or a loss depending on one's viewpoint. From the perspective of a pine beauty moth in Scotland, a new commercial plantation of lodgepole pine from western North America creates a much better habitat than the moorland it replaces, although it brings a catastrophic loss to many native species of wetland plants, insects and birds. Even when there is a narrower focus on a few species of particular interest, the position may not be clear. Is it a gain or a loss when conservation activity leads to an increase in the population of the (endangered) pine marten which is then found to be preying on (endangered) capercaillie?[28] Nevertheless, the paradigm case will involve the transformation of an area of (semi-)natural habitat into land dedicated to intensive agriculture, infrastructure, urban uses or mining, so that the loss is clear.

The obvious way to compensate for such a loss would be to achieve on a nearby site the reverse conversion from industrialised or built-up land to valuable habitat, but this may not be practicable for technical reasons even in the unlikely event of appropriate land being made available. An alternative approach is to look for a gain not in providing new or enhanced habitat but in removing a threat to existing habitat. If a site currently offering biodiversity value is on its way to being destroyed or degraded, then providing effective protection for it can be argued to be achieving a benefit; in the face of the relentless decline in biodiversity, even maintaining the status quo can be seen as a gain. The strength of this argument depends partly on how realistic the threat is to the site that is now being conserved,[29] and the potential for other conservation measures. Although there may well be an improvement on what would have been the position without this intervention, nevertheless it is harder to justify such measures as gains for the purposes of an offset if the standard is that of achieving no net loss to biodiversity.[30] In between

---

[28] Fiona Matthews, 'Pine marten-capercaillie conflict: should pine martens in Scotland be culled?', *Mammal News*, Autumn 2012, available at https://biosciences.exeter.ac.uk/documents/MammalNewsAutumn2012.pdf.

[29] When the UK system of Sites of Special Scientific Interest allowed for payments to landowners on a profits foregone basis, compensating for the profits that would have been made had proposals to convert sites to more commercial uses not been stopped, it was argued that several proposals were purely speculative and put forward mainly to attract the payments for abandoning them; Colin T. Reid, *Nature Conservation Law* (1st edn, W. Green, 1994) 34.

[30] The New South Wales BioBanking scheme requires offsets to 'improve or maintain biodiversity values' which can include protecting existing areas of

these examples is the possibility of enhancing habitat that has not been completely destroyed but is degraded and offering less than its potential in supporting biodiversity. Here it is easier to see that there has been a gain, and the greater difficulty lies in quantifying it.

A significant issue in all of these scenarios may be the lack of good baseline data. In many cases, and particularly for invertebrates and lower plants, there will be a lack of the basic information on which assessments of biodiversity gains and losses can be made.[31] In the EU examples given earlier, the problem is avoided by applying the compensatory measures in contexts where the loss is defined in terms of the condition of designated areas or species. In such cases there should be baseline data already to hand detailing the previous state of the environment, prepared for the designation of habitats and species and their monitoring as required by the law. In other contexts this assessment will not be so straightforward, but being able to establish the biodiversity gains and losses with an adequate degree of certainty is an essential building block of any officially recognised offsetting scheme.

4.1.1 Additionality

A separate issue in identifying any biodiversity gain is that of additionality.[32] In order to recognise any conservation activities as providing a gain specifically to offset a particular loss, they must be shown to be activities that would not otherwise be occurring. If no net loss to biodiversity is to be achieved, then losses must be balanced by something additional, over and above what would be occurring in the ordinary run of business. Various difficulties may arise at this stage. In the first place, 'business as usual' may present a moving target; what is viewed as the normal practice changes. In European agriculture, for example, some of the features which are now accepted as minimum standards for any financial support not so long ago might have attracted special payments,[33] so that identifying what is additional, and above all specifying

---

value; Department of Environment and Climate Change NSW, *BioBanking Scheme Overview* (2007), available at www.environment.nsw.gov.au/resources/biobanking/biobankingoverview07528.pdf. Under the US wetland mitigation scheme, an offset provided through preserving habitat must secure more land than is being adversely affected; 33 CFR §332.3(f)(2), 40 CFR §230.93(f)(2).

[31] See Chapter 2, section 2.11.

[32] See Chapter 2, section 2.5.

[33] Compare the requirements in the early orders designating Environmentally Sensitive Areas under the Agriculture Act 1986, s. 1 (e.g. Environmentally Sensitive

this at the beginning of what may be a long-term project, may be difficult. Secondly, the offset activity may not be occurring in isolation. If developers meet their offset commitments by providing financial support for activities, rather than taking operational responsibility themselves,[34] the funds provided may very well not be the only contribution to a conservation project. It is common for projects to draw in funds from various sources. In some cases these may allow for incremental growth, so that a distinct proportion of the activity can be attributed to each source of funding. In other cases, though, the arrangements will be more integrated, with each contribution being vital in determining whether or not the project can proceed at all (such as the total being needed to provide the purchase price of the site in question, or specific contributions being used for essential preparatory work or to provide the 'match-funding' without which other forms of support are not available).[35] In such cases it would be misleading to attribute the whole of the biodiversity gains to the offset arrangement, even though without that piece in the jigsaw the project could not have proceeded. In such circumstances, measuring how much gain can be attributed to the offset contribution may not be straightforward.

Related concerns have been raised by a study in Australia considering potential perverse consequences of offset arrangements.[36] This raises fears that offset projects may absorb some of the volunteer and other resources which are currently devoted to conservation, so that rather than such efforts contributing to genuine biodiversity gains, these are expended in activities which are merely seeking to ensure no net loss. Conservation effort might thus be diverted into providing what is in effect a subsidy to the developer who should be paying for all the

---

Areas (Breadalbane) Designation Order 1987, SI 1987/653) with those for ensuring 'good agricultural and environmental condition' under the cross-compliance rules as a prerequisite of any support under the Rural Development Programme (Scotland) (see Scottish Government, *Rural Payments and Services* (2015), available at www.ruralpayments.org/publicsite/futures/topics/inspections/all-inspections/cross-compliance/detailed-guidance/good-agricultural-and-environmental-conditions/.

[34] See section 4.4 below.

[35] Scottish Borders Council has said that in operating its offsetting scheme (in reality an impact fee rather than a true offset; see the end of Chapter 7, section 7.4) the developers' contributions have been key to attracting through various schemes run by public and charitable bodies total funding of up to 10 times the amount of those contributions for biodiversity projects in their area (personal communication).

[36] Ascelin Gordon and others, 'Perverse incentives risk undermining biodiversity offset policies' (2015) 52 *Journal of Applied Ecology* 532.

measures to offset the biodiversity losses being caused. A further fear is that an apparently functioning offset market might lead government to reduce other conservation efforts and create a false sense of confidence that nature is being safely looked after, despite the risks in any offset schemes.

4.2 PRACTICALITY AND EFFECTIVENESS

In contemplating the potential for compensating for biodiversity losses by gains elsewhere, a further issue arises, before addressing the details of equivalence, monitoring and guarantees, namely whether the gains proposed from the creation or enhancement of habitat can actually be achieved in practice.[37] The first challenge is finding a suitable site for undertaking the offsetting activity, a challenge that becomes more significant the narrower the range of offsets deemed to be acceptable. If a scheme demands an offset that provides exactly the same sort of habitat, or measures taken in the close vicinity of the loss,[38] then the range of alternative sites for the offsetting activity is significantly constrained. There are both ecological and 'socio-economic' aspects to this.

At the ecological level, there may be great difficulty in finding an alternative site where it is possible to provide the equivalent of what is being lost; as discussed below, this depends on how far the compensatory habitat must match in kind what is being lost. Each habitat has its own requirements in terms of variables such as geology and soil chemistry, temperature, elevation, rainfall and water availability, so that if there is any requirement for similarity between what is lost and what is gained there will be a limited supply of potential sites for compensatory action. At its simplest, any attempt to replace a coastal habitat can only take place at another coastal site and a lot of the coast will have quite different features or have already been built on in a way which physically precludes a 'return to nature'. There simply may not be suitable sites where the requisite conditions can be provided for the features that are being lost to be replaced. This is especially true for habitats which rely on physiographical features with a limited distribution. Such sites may be regarded as especially valuable because of their rarity, and the fact that

[37] This section focuses on the achievement of functional success, ignoring the further question of whether a habitat restored or recreated through human intervention can ever have the same value as a truly authentic one, as explored in Chapter 8, section 8.2.

[38] See section 4.3 below, text to nn. 66 and 69.

no direct substitute can be provided may provide an argument for excluding them from any offsetting scheme.[39]

Even if ecologically suitable sites exist, there is the question of their availability. This raises a number of socio-economic issues, resting on the fact that, especially in densely populated areas, there is unlikely to be 'spare' land available, waiting to be dedicated to conservation purposes. Much will depend on the patterns of land ownership and use. Where there are substantial tracts of land in state ownership and not in commercial use, then allocating land for the purposes of an official offsetting scheme may be more straightforward. Yet in a heavily populated and heavily developed country like the United Kingdom, where most land is in private ownership and devoted to some productive use, finding any substantial site to use for conservation may be difficult. This will be especially so for the less widely distributed habitats which rely on features which have also proved attractive for human uses, such as estuaries, and in the regions where the greatest development pressure, and thus the greatest need for offsets, is to be found. The problems may be eased if other mechanisms, such as payments for ecosystem services or under a conservation covenant mean that conservation itself becomes an income-producing use of land,[40] but there is still the need to find landowners willing to sell their land or to commit it in some other way to the conservation use.[41] The availability of compulsory purchase powers may ease the position,[42] but experience is that following that route can often be contentious, costly and time-consuming.[43] Finding a location for delivering the conservation gains needed for an offset may thus be a problem in itself, and of course dedicating the new site to create the

---

39 See section 4.3 below, text to n. 65.

40 See Chapters 3 and 5.

41 Thomas Kabii and Pierre Horwitz, 'A review of landholder motivations and determinants for participation in conservation covenanting programmes' (2006) 33 *Environmental Conservation* 11.

42 Such powers are often available in conjunction with major infrastructure and development projects, and for many other purposes seen as being in the public interest; see e.g., the long and varied list of legislation authorising compulsory purchase in Scotland: Scottish Law Commission, *Discussion Paper on Compulsory Purchase* (2014) 331–8.

43 E.g. the litigation arising from the compulsory acquisition of land to provide new inter-tidal habitat to compensate for the loss of mudflats as a result of the construction of the Cardiff Bay Barrage; *Waters* v. *Welsh Development Agency* [2004] UKHL 19, [2004] 1 WLR 1304.

offset may involve further losses of various sorts that have to be taken into account.[44]

Actually providing the gain poses its own problems. It is still comparatively early in our experience of translocating species and restoring and (re-)creating habitats and it is not clear how successfully this can be done, especially in the case of those habitats which take decades, or centuries, to mature, such as woodland. Indeed, an underlying problem may be in determining what is to count as success[45] and when this has been attained:

> [B]ecause restoration is often a long-term process, deciding when a project has 'succeeded' is challenging. In the case of relatively narrow objectives, such as re-introduction of a species, or elimination of an invasive species, targets may be set, but this is harder for more general ecosystem-scale restoration.[46]

Restoration activities can achieve gains[47] but often fall short of their objectives,[48] and even in habitat which responds quickly, the results will not have the same biological characteristics as naturally occurring habitat.[49] Objectives set in terms of achieving equivalent ecosystem

---

44 Losses may include the loss of current biodiversity, especially if land is converted to provide a new form of habitat such as grassland becoming woodland, or the loss of the different potential biodiversity that could be achieved if the site were managed to promote features other than those which are the focus of the offset. Other losses might include the loss of actual or potential capacity to deliver other human objectives, such as agriculture productivity or recreation, and the loss of authenticity and true 'wildness' if a site which has previously seen little human intervention becomes subject to more intrusive management, even in the name of 'nature' (see Chapter 8, section 8.2).

45 Under the US wetland mitigation scheme, performance standards must be set related to the conservation objectives so that there can be an objective evaluation of whether the required objectives are being met; 33 CFR §332.5(a), 40 CFR §230.95.

46 Karen A. Keenleyside and others, *Ecological Restoration for Protected Areas: Principles, Guidelines and Best Practices* (IUCN, 2012) 70.

47 See e.g., the case studies in *ibid.*

48 Martine Maron and others, 'Faustian bargains? Restoration realities in the context of biodiversity offset policies' (2012) 155 *Biological Conservation* 141.

49 Hannah L. Mossman, Anthony J. Davy and Alastair Grant, 'Does managed coastal realignment create saltmarshes with "equivalent biological characteristics" to natural reference sites?' (2012) 49 *Journal of Applied Ecology* 1446. This study notes that aiming at equivalence in ecosystem function, rather than exact ecological equivalence, may be more feasible.

functions rather than ecological equivalence may be more attainable,[50] but even here the time-scales can be very substantial, for example, one study of how arable land in England might be returned to floodplain meadows suggests that it will take over 150 years for all the appropriate species to become established (and even then not fully matching ancient meadows) and over 60 years for the functional traits of the plant community to be replicated.[51]

There are many variables to be taken into account if success is to be achieved and one review identified poor definition and measurability in what is to be achieved, uncertainty in achieving the outcomes and the time-lags before goals can be reached as main factors limiting the success of offset projects.[52] What is clear, though, is that the process is not straightforward, with ecological, genetic and human factors to be considered,[53] and that success is far from guaranteed.[54] Identifying and investing in 'insurance' sites in case the main offset fails may thus be appropriate,[55] although this would obviously increase the cost and lower the attractiveness of an offsetting project. A long-term commitment, careful monitoring of progress and the adoption of adaptive management, reacting to progress on the ground, are important,[56] whilst the needs of each species and habitat must be considered, with different approaches being needed in different situations, and different features being more or less amenable to restoration attempts.[57] As is stated in the conclusion of

---

50 *Ibid.* 1454.

51 Ben A. Woodcock, Alison W. McDonald and Richard F. Pywell, 'Can long-term floodplain meadow recreation replicate species composition and functional characteristics of target grasslands?' (2011) 48 *Journal of Applied Ecology* 1070.

52 Maron and others, 'Faustian bargains?' (n. 48).

53 Kathy H. Hodder and James M. Bullock, 'Translocations of native species in the UK: implications for biodiversity' (1997) 34 *Journal of Applied Ecology* 547.

54 Tammy Hill, Eric Kulz, Breda Munoz and John R. Dorney, 'Compensatory stream and wetland mitigation in North Carolina: an evaluation of regulatory success' (2013) 51 *Environmental Management* 1077.

55 Rebecca Kormos and others, 'Great apes and biodiversity offset projects in Africa: the case for national offset strategies' (2014) 9(11) *PLOS One e111671* 12.

56 Keenleyside and others, *Ecological Restoration for Protected Areas* (n. 46).

57 Karin Johst and others, 'Biodiversity conservation in dynamic landscapes: trade-offs between number, connectivity and turnover of habitat patches' (2011) 48 *Journal of Applied Ecology* 1227, 1233–4.

a study on a project to relocate the green and golden bell frog from sites within Sydney's Olympic Park:

> [H]abitat offset aimed at achieving and detecting no net loss can only be successful where the offset ratio is large, monitoring is long-term, robust and precise and funding is available to substantially increase the amount of habitat if monitoring indicates that this is necessary. This is the major short-fall of most offset programs, and this paper illustrates that even for species that are perceivably ideal for habitat offset, a large amount of effort is required for successful outcomes.[58]

Such conclusions do not rule out a role for biodiversity offsetting, especially if the realistic alternative is simply losing biodiversity with no compensatory measures at all, but should be borne in mind in deciding on the details and scope of any scheme.

4.3 EQUIVALENCE AND EXCHANGEABILITY

Any offsetting approach relies on there being an acceptance of exchangeability between what is being lost and what is being gained and a means of assessing their equivalence. In some environmental contexts, such issues present fewer difficulties than for biodiversity. The main concern in relation to greenhouse gases is the concentration of these gases in the global atmosphere and it is possible to develop a common unit of exchange, whereby different substances can all be given a value in the one currency providing a single way of measuring the value of different things. Emissions involving a number of gases are all included on the basis of carbon(-dioxide) equivalents,[59] enabling a number of greenhouse gases to be taken into account, recognising their different contributions to climate change.[60] The fact that greenhouse gas schemes have concentrated on a single aspect of the emissions makes the task of establishing a common currency more straightforward, but these schemes share

[58] Evan J. Pickett and others, 'Achieving no net loss in habitat offset of a threatened frog required high offset ratio and intensive monitoring' (2013) 157 *Biological Conservation* 156, 161.

[59] R.K. Pachauri and A. Reisinger (eds), *Climate Change 2007: Synthesis Report, Contribution of Working Groups I, II and III to the Fourth Assessment Report of the Intergovernmental Panel on Climate Change*, (IPCC, 2007) s. 2.1.

[60] Methane makes a contribution per tonne 25 times that of carbon-dioxide; S. Solomon and others (eds), *Contribution of Working Group I to the Fourth Assessment Report of the Intergovernmental Panel on Climate Change* (CUP, 2007) Table 2.14.

another feature which marks them out from biodiversity examples, the essentially fungible nature of the emissions being considered.[61] Although there are also very localised impacts to be considered in controlling emissions,[62] what really counts in this context is the overall state of the atmosphere on a large scale. This enables an approach to be taken that can easily accommodate trade-offs between different sorts of emissions at different locations so long as the overall position remains within the limits set at a broad level.

For biodiversity, things are very different. Every site is unique in its particular set of environmental conditions and its species mix and these cannot be exactly reproduced elsewhere. This uniqueness is becoming ever more apparent as we study the micro-fauna and flora more thoroughly and our understanding of genetics reveals the variety within and between local populations of the same species. Offsetting therefore depends on accepting that the loss of a site (or some of its features) can be compensated for by providing something not identical but which is deemed to be of equivalent value. It may seem wrong to indulge in a comparative valuation of different aspects of biodiversity, but it must be recognised that such assessment of relative value is already embedded in the more traditional approaches to conservation. The designation of habitats and species for protection, a mainstay of conventional approaches, rests on making judgements about the relative importance of different elements and on the identification of those which merit strict protection and those which do not merit such special consideration. What is different here is the attempt to create a more structured way of making comparisons, potentially even creating a currency, which can be applied to varied circumstances.

There are many factors that can influence such a comparative framework and many choices to be made in determining equivalences.[63] A scheme that requires the offset provided to match closely what is being lost may appear to be the strongest in conserving the existing mix of

---

61 Salzman and Ruhl correctly refer to the 'range of fungibilities' exhibited by the commodities involved in environmental traded markets; James Salzman and J.B. Ruhl, 'Currencies and commodification of environmental law' (2000–01) 53 *Stanford Law Review* 607, 612.

62 Very large emissions of greenhouse gases in one place create an immediate risk to the health of those in the vicinity, not just a contribution to climate change at a global level.

63 Metrics used in Australia, England, France, Germany, South Africa and the United States are reviewed in Annex 1 to ICF Consulting Services for the European Commission, *Final Report* (n. 9).

biodiversity, but this will seriously limit the flexibility of the scheme and miss the potential to take a wider view and use offsetting to bring benefits for biodiversity beyond a narrow replacement (as far as possible) for what is being lost. Locational issues must be considered for both ecological and human reasons, whilst scale and time are also relevant. In any event, allowance should be made for the fact that success in any project aimed at achieving biodiversity gains is not guaranteed, a fact that underlies criticism by the Office of the Inspector General of the US Environmental Protection Agency for not making it clear that its reporting of 'no net loss' in wetland mitigation was based on an assumption of complete success of work done despite studies showing that as many as one-quarter of the projects might not meet their performance standards.[64]

In considering such matters, it must be stressed that some areas or features may be recognised as so valuable in themselves that no biodiversity loss, however fully compensated, is acceptable. This can be done by the prior exclusion of certain areas from any offsetting scheme (for example, limiting the scope of offsetting to areas outwith certain categories of protected habitats) or may be the result of initial calculations revealing that offsetting should not be followed for the site because of the scale or nature of what is being lost. Under the New South Wales BioBanking scheme, there are 'Red Flag' areas identified on the basis of landscape features, native vegetation and threatened species and within these many impacts are excluded from the scope of the offset arrangements, unless express, site-specific approval is granted.[65] Accepting the existence and operation of an offsetting scheme does not commit one to accepting that every feature of the ecosystem is 'up for grabs' and vulnerable to damage so long as an offset can be offered. Offsets are best suited to being a supplement to, not a substitute for, more direct conservation measures.

4.3.1 Replacement or Substitution?

One approach to offsetting schemes is to see their purpose as providing a direct replacement for what is being lost. On this basis the only sort of

---

[64] Office of Inspector General, *EPA Needs to Clarify Its Claim of 'No Net Loss' of Wetlands*, Report No. 14-P-0191 (2014), available at www.epa.gov/office-inspector-general/report-epa-needs-clarify-its-claim-no-net-loss-wetlands.

[65] New South Wales Government, *BioBanking Assessment Methodology 2014* (Office of Environment and Heritage, 2014) 31–9. 'No-go zones' are also recommended by Kormos and others, 'Great apes and biodiversity offset projects in Africa' (n. 55).

compensatory action which is acceptable is that which will provide a 'like-for-like' offset; if saltmarshes are being lost, then saltmarshes must be provided. As noted above, however, no one site can be identical to another, and therefore even here there are choices to be made in how precisely the habitat to be replaced is described, what features are to be counted in assessing equivalence and whether the emphasis is to be on species mix or ecological functions.[66] Insisting on a like-for-like offset inevitably constrains the scheme to a great extent, especially since, as noted above, there may be practical difficulties in finding, or making use of, suitable alternative sites. It simply may not be possible to provide a direct equivalent. Allowing a substitution of a different sort of habitat, but of equivalent value, is more in keeping with the market-based, trading ideas which underlie offsetting, and also presents opportunities to achieve not just no net loss to biodiversity but net gains.

Such shifting between different elements of biodiversity can have positive or negative impacts when the wider picture is considered. By allowing an element of 'trading up', the loss of sites which are of no special significance can be the trigger for action that will benefit those of greater importance, because of their rarity, endangered status or quality in biodiversity terms.[67] In this way the effort and expenditure generated by the requirement for an offset can be directed so that features which are of great significance can benefit at the expense of sites of no outstanding value. On the other hand, if the same logic is followed without constraints, then very special sites might be lost and the equivalent value provided by larger areas of less special habitat, diminishing the overall diversity that exists. Moreover, there is a danger of offset provision concentrating on the forms which are easiest and cheapest to provide, at the expense of the more challenging offerings, skewing the overall balance of diversity in the areas affected. Such dangers can be addressed in various ways, such as allowing 'trading up' but not 'trading down', in

---

[66] Mossman, Davy and Grant, 'Does managed coastal realignment create saltmarshes' (n. 49) 1451–4. The US wetland mitigation scheme endeavours to compensate for 'the aquatic resource functions' that will be lost: 33 CFR §332.3(a)(1), 40 CFR §230.93(a). The New South Wales BioBanking scheme operates on the basis of matching credit profiles based on the vegetation type under the system of classifying native vegetation approved by the NSW Plant Community Type Control Panel and the sub-regions under the Interim Biogeographic Regionalisation for Australia (IBRA) system; New South Wales Government, *BioBanking Assessment Methodology 2014* (n. 65) 43.

[67] Of course, the criteria set to determine the relative value of different species, sites or habitats may themselves be contestable.

other words allowing offsets to provide for habitat of more but not less significance, or incentivising offset provision which contributes to the approved conservation strategy.[68]

4.3.2 Location

The location of both the site being lost and the offset that is being provided are relevant for several reasons. In the first place, location is one of the key factors in determining the nature of the habitat or species that are being lost or can be provided on an offset site. An upland site cannot be a direct replacement for a coastal one, and the same applies with relation to many more subtle variables, such as soil chemistry, water levels and local climate. Location is thus a key consideration that affects how closely an offset can replicate exactly what is being lost.[69]

Secondly, a site's location may significantly affect its biodiversity value. An area of woodland when assessed in isolation may not offer any great biodiversity value, yet it may have a vital role to play if it is linking other areas of greater inherent value. Building on a site which has a part to play in a network of 'green corridors', or provides a flyway for bats between their roosting and foraging grounds, may result in losses far greater than seem likely if the site is assessed on its own. Losing an area of similar quality in another location may be a trivial loss, but the connections with other features can transform the impact of the deterioration of any site. The ecological value of the offset being provided will also vary according to location, with sites adjacent to or connected with other sites with healthy biodiversity likely to offer more benefit than those in isolation.

A third consideration is a human one. An area of peri-urban woodland or meadow, subject to disturbance and with a degraded range of species present, may not be of any great ecological value. In pure ecological terms, therefore, providing a substitute in another region, connected to richer sites and in an area where wildlife can flourish, is likely to offer much more by way of biodiversity benefit. Yet this takes no account of the value to local human residents. The site may not be of great ecological value, but it may be very important in terms of amenity and of

---

[68] As in the pilot scheme in England discussed in section 4.3.5 below.

[69] The US wetland mitigation scheme expressly aims for compensation within the same watershed; 33 CFR §332.3(b)(1), 40 CFR §230.93(b); Philip Womble and Martin Doyle, 'The geography of trading ecosystem services: a case study of wetland and stream compensatory mitigation markets' (2012) 36 *Harvard Environmental Law Review* 229.

providing residents with access to nature. An offset site many miles away may offer equivalent or greater value in terms of flora and fauna, but offers nothing to make up for the loss of its contribution in other ways. The potential for offset schemes to produce such a shift in where biodiversity sites are located has been shown by studies of wetland mitigation in the United States, which report a 'migration of wetlands and streams from higher population density, urban or suburban areas to low population density, rural areas'.[70]

4.3.3 Coherence

In offsetting schemes for greenhouse gases, environmental gains can be achieved by any reduction in emissions, and each project can be looked at in isolation since each makes a valuable contribution to the global atmospheric concentrations, independently of what is happening elsewhere. For biodiversity, it is different. The value of any site and the benefits it can bring for biodiversity are often dependent on what is happening elsewhere and a coherent conservation strategy must be pursued if benefits are to be achieved. This need for coherence arises both from what is required to ensure the health of particular sites and from the needs of particular species. Regardless of the effort put into enhancing the site itself, a wetland cannot survive unless both the upstream and downstream conditions and water flow are also maintained, whilst migratory species can thrive only if there are suitable conditions available to support all stages of their annual cycle, so that isolated measures to conserve or enhance their breeding sites are of no use unless wintering and transit sites are also maintained.

It follows, therefore, that the value of any loss or gain must be assessed not in isolation but in the context of what is being contributed to wider conservation plans. This in turn is likely to require the production of explicit conservation strategies to guide the offset market. A clear connection exists between this point and those discussed above, since one pitfall to be avoided in introducing an offsetting scheme is the loss of a wide range of biodiversity (in type and location) to be replaced by offsets that concentrate on particular species or geographical areas. Given the development pressures around cities and the relative cost of land, it is possible to foresee the export of biodiversity from the areas of economic growth to the remoter and less-developed regions, skewing the balance of

---

70 *Ibid.* 274.

biodiversity being encouraged and rendering the developed areas 'nature-free zones'. In a system based largely on a network of officially designated conservation sites, coherence is essentially a matter of internal policy for the relevant conservation authorities, but when the distribution of conservation sites is left to the market, ensuring an overall pattern which can deliver effective conservation and enhancement across the range of biodiversity requires a different approach. Indeed, this is a further argument which may point towards some sites requiring absolute protection rather than being 'in play' under an offsetting scheme.

4.3.4 Timing

Temporal considerations are another factor to be taken into account in assessing the equivalence of offsets. Whereas the loss of biodiversity when a site is developed in any way is likely to be almost immediate, the compensating gains may take a long time to be achieved.[71] The question therefore is whether an offsetting gain should be recognised when steps are taken to provide biodiversity benefits or only when the benefits are actually being delivered. Except in cases where conservation of existing habitat is accepted as a gain, there will always be some delay, and possibly a very considerable delay,[72] before enhancement or restoration actions produce results. Indeed, given the certainty of the habitat loss and the lack of certainty over its successful replacement, one might argue that a strict application of the precautionary principle should rule out accepting any offsets except with retrospective effect. The significance of this issue in the design of any scheme is increased if offsets are restricted to like-for-like provision and the areas being lost contain habitat that takes a long time to establish.[73] In such cases, an offset is offering a distant (and inevitably not wholly guaranteed) gain in exchange for an immediate loss, which entails at best a temporary dip below the level of no net loss to biodiversity. A requirement for interim measures to provide biodiversity benefits whilst the longer-term arrangements are maturing is

---

[71] R.K.A. Morris and others, 'The creation of compensatory habitat: can it secure sustainable development?' (2006) 14 *Journal for Nature Conservation* 106.

[72] As shown in the example discussed in Woodcock, McDonald and Pywell, 'Can long-term floodplain meadow recreation replicate species' (n. 51).

[73] In Germany, only compensatory measures that will deliver results within 25 years are accepted in practice; Albrecht, Schumacher and Wende, 'The German impact-mitigation regulation' (n. 8) 319.

one possible response to this (as under the EU Environmental Liability Directive).[74]

4.3.5 Finding a Metric

Finding a way to reconcile these different considerations is challenging and can be done with different levels of precision; 'the metric must be sophisticated enough to reflect the biodiversity value of development sites while remaining transparent and user-friendly but without leading to perverse incentives'.[75] A range of scientific assessments and methodologies are being tested and adopted[76] and one straightforward attempt to identify and cater for these variables is provided in the Guidance produced for the pilot scheme for biodiversity offsetting developed in England by the Department for the Environment and Rural Affairs (DEFRA).[77] This places the emphasis on producing a method that is simple to operate,[78] so that all parties involved can ascertain quickly[79] and clearly what needs to be provided, assisting both those on the demand site (the developers or landowners whose activities are generating the need for an offset) and those on the supply side (such as landowners or conservation groups in a position to offer sites where the biodiversity losses can be offset). Nevertheless, it does attempt to take

---

[74] Known as 'compensatory remediation', Annex II to Directive 2004/35/CE.

[75] House of Commons Environmental Audit Committee, *Biodiversity Offsetting: Government Response to the Committee's Sixth Report of Session 2013–14*, Ninth Special Report of Session 2013–14, HC 1195, para. 8.

[76] See eftec, IEEP and others, *The Use of Market-based Instruments for Biodiversity Protection: The Case of Habitat Banking, Technical Report for European Commission DG Environment* (eftec, 2010) 51–61, 162–8.

[77] Department for Environment and Rural Affairs, *Biodiversity Offsetting Pilots: Guidance for Developers* and *Guidance for Offset Providers* (2012). For further background on this pilot see Jo Treweek, *Scoping Study for the Design and Use of Biodiversity Offsets in an English Context, Final Report to Defra*, Contract NE 0801 (2009).

[78] This scheme is very simple compared to some others, e.g. completion of a four-day training course on the BioBanking Credit Calculator is needed to become an assessor under the New South Wales BioBanking scheme: New South Wales Government, *Credit Calculator for Major Projects and BioBanking* (2014), available at www.environment.nsw.gov.au/biobanking/calculator.htm.

[79] One government suggestion was that the metric could take 'as little as 20 minutes' to apply to a site; DEFRA, *Biodiversity Offsetting in England: Green Paper* (DEFRA, 2013) para. 25.

into account the range of issues discussed above,[80] albeit by making use of broad categories rather than precise and individualised assessments.

The proposal is for a calculation of the biodiversity units being damaged or generated based on a number of variables. In determining the harm to be offset, account should be taken of three factors:

- the area of land concerned;
- the distinctiveness of the habitat being affected, measured according to a list of habitats and their distinctiveness ratings, on a three-point scale, given in an Appendix to the Guidance;[81]
- the existing condition of that habitat, again according to a three-point scale,[82] namely poor, moderate or good based on a standard framework.[83]

When multiplied together, these produce a figure for the number of biodiversity units that will be lost on a site and that therefore must be provided by the offset being sought. In addition to the figures generated by those steps, two further elements relate to the sort of habitat that can be provided through the offsetting process. This does not limit the scheme to like-for-like offsets but does impose some constraints based on habitat type and does prevent 'trading down'. First, the loss of habitat of high distinctiveness must be replaced with other habitat of similar distinctiveness and usually of the same type. Secondly, habitat in medium or poor condition can be replaced only by habitat in high or medium condition.

The same elements are used to determine the value of any offset being provided, but three additional factors are also to be taken into account, operating as discounting rates,[84] reducing the value initially determined

---

[80] Extreme cases are not addressed since the scheme is designed to offset residual losses which cannot be avoided or mitigated on site, and is not to be applied to particularly important or sensitive sites; the planning hierarchy requires that proposals damaging such sites should just be refused permission rather than accommodated by means of offsetting (see n. 19).

[81] High distinctiveness produces a multiplier of 6, Medium 4 and Low 2.

[82] Good condition produces a multiplier of 3, Moderate 2 and Poor 1.

[83] In the pilots this was based on the Manual used by Natural England in determining support under agri-environment schemes: Natural England, *Higher Level Stewardship Farm Environment Plan (FEP) Manual* (3rd edn, Natural England, 2010).

[84] Under the US wetland mitigation scheme there must be a ratio of greater than one-to-one between the areas lost and gained where this is necessary to account for the method of conservation, the difficulty and likelihood of success,

by the calculation based on area, distinctiveness and condition. These discounting factors are:

- the difficulty of restoring or recreating the habitat in question, on a four-point scale from Very High to Low;[85]
- the location of the offset site, with the discount applied when sites do not fit with wider strategies for biodiversity offsetting;[86]
- the length of time before the replacement habitat will be in mature condition, using a 3.5 per cent discount rate.[87]

The metric therefore does endeavour to take account of the various factors which affect value and equivalence, and presents them in a way which provides a simple and clear means of calculating what is being lost and gained. This simplicity, however, has been the focus of considerable criticism. The metric has been described as 'overly simplistic', with a call for a metric that will instead 'reflect the full complexity of habitats'.[88] The telling point was made that, 'Having a system that looks very simple is actually quite hard to operate. You end up making a lot of value judgments in there, because there are not enough categories to choose from.'[89] This specific proposal may have erred too far in the direction of simplicity, but in terms of controlling transaction costs there are clear advantages in having some standardisation in the methodology rather than each assessment having to be done on a wholly individualised basis. Finding a balance that is widely acceptable may not be straightforward,[90]

---

differences between the functions lost and gained and the distance between the development and the offset site; 33 CFR §332.3(f)(2), 40 CFR §230.93(f)(2).

[85] Very High difficulty produces a multiplier of 10, High 3, Medium 1.5 and Low 1.

[86] There is no discount when the location is identified in the offsetting strategy, a multiplier of 2 when it is buffering, linking, restoring or expanding a habitat outside an area identified in the strategy, and a multiplier of 3 if there is no contribution to the strategy.

[87] Accordingly, if it will take five years to reach the target condition, a multiplier of 1.2 applies, for 20 years 2.0 and 30 years 2.8.

[88] House of Commons Environmental Audit Committee, *Biodiversity Offsetting*, Sixth Report of 2013–14, HC 750, para. 16.

[89] Evidence from the Environment Bank; *ibid.* para. 10.

[90] '[C]orrecting one source of inefficiency (e.g. lack of flexibility) may result in loss of other benefits, such as upfront certainty and objectivity': New South Wales Government, *BioBanking Scheme: Statutory Review Report* (Office of Environment and Heritage, 2014) 14, available at www.environment.nsw.gov.au/biobanking/bbfindings.htm.

and the more sophisticated and individualised the metric, the higher the transaction costs in reaching any particular offsetting arrangement.

4.3.6 Disaggregation

A different response to deal with some of the complexity is to contemplate a disaggregation or 'unbundling' of the elements of biodiversity that are being lost and gained. A site that is to suffer biodiversity losses as a result of development will offer a range of ecosystem services and rather than trying to find a single offset site which can supply all of these, it might be easier and more practical to disaggregate these and attempt to provide equivalent services across several sites. Thus, an area of urban woodland that is of value to local residents for amenity purposes and as their link with nature, as well as being important for various woodland species, might be replaced by two different sites, one near at hand which will provide the amenity for residents, even though the ecological value of the new site may not be great, and a second which can seek to maximise the ecological benefits, for example, through being adjacent to another area of woodland of significant biodiversity value, even though this may be a substantial distance from the original site. Where there are further ecosystem services, such as an impact on local drainage, then further measures might be required to address these, but doing so directly, concentrating solely on what can be done to remedy that one impact, rather than having to incorporate ecological and amenity benefits at the same time. Such an approach does perhaps require an even more sophisticated analysis of the biodiversity and ecosystem services that are being lost, and can be gained, but may be more practical and effective than trying to provide in one new location all the attributes of a site that are being lost. There are obvious similarities here with the process of identifying and dealing separately with various ecosystem services for the purposes of payment, as discussed in Chapter 3, section 3.2.

4.4 PARTIES AND RELATIONSHIPS

Since an offset is essentially a simple trade to balance a loss with a compensatory gain, it may seem that the relationships involved will be simple, taking the form of a straightforward deal between the party causing the loss and seeking a gain to compensate (the 'purchaser') and the one supplying that gain (the 'provider'). In practice though, there is likely to be a more complex set of relationships with more parties involved.[91] There are two dimensions to this, the mechanics of providing the gain and the presence of some element of external approval or verification.

Before considering such practicalities, though, there is the question of when the offsetting process is going to be used. A key issue here is whether any scheme is to be voluntary or mandatory. For a mandatory scheme, for example, requiring offsets when development permits are sought for a project that causes any biodiversity loss (or losses over a certain threshold), such a requirement produces a guaranteed and substantial number of parties seeking to acquire offsets. This demand will stimulate the providers who are in a position to offer offset arrangements to do so, knowing that there is a market for this service. Moreover, the mandatory nature of the scheme will increase the resources and effort being directed toward conservation. The scale of business should also lead to lower transaction costs, with greater experience and efficiencies of scale.[92] On the other hand, monitoring and regulation may have to be stricter since not everyone will be a willing participant who is committed to the goal of avoiding biodiversity losses. There will also be greater strength in the market forces that tend towards the cheaper and simpler forms of offsetting, potentially distorting the focus of the conservation effort. By contrast, if a scheme is voluntary, the very fact of voluntary engagement may justify a lighter touch in regulation (although it must be remembered that offset arrangements may have to endure long after the original willing participants are no longer around). However, it may be that incentives can be provided, for example, through subsidies or tax benefits, to encourage participation beyond a few enthusiasts.[93] The small size of a fully voluntary market, even with incentives, is unlikely to

---

[91] Moreover, a single scheme can allow for various options, such as the US wetland mitigation scheme which provides for permittee-responsible mitigation (where the developer takes direct responsibility), mitigation banking and in-lieu fees; 33 CFR §332.3(b), 40 CFR§230.93(b).

[92] Finding a suitable partner was one aspect of the transaction costs identified in studies in Australia; Anthea Coggan and others, 'Transferable mitigation of environmental impacts of development: two cases of offsets in Australia' (2013) 15 *Journal of Environmental Policy and Planning* 303, 316.

[93] An example that perhaps sits between the mandatory and voluntary extremes is the New South Wales BioBanking scheme, where although not the only way of dealing with potential impacts, having an approved bio-banking statement is conclusive evidence that a project is not likely to significantly affect any threatened species, population or ecological community, overcoming one hurdle which stands in the way of planning approval; Threatened Species Conservation Act 1995, s. 127ZO.

encourage the scale of activity that will generate much interest or stimulate potential offset providers to see this as a viable way of using their land. The scale of the market generated by mandatory requirements is a powerful argument for having such an element if offsetting is to become a significant element of conservation policy.[94]

Turning to the mechanics of providing the offset, there are several ways in which the purchaser and provider can go about this task. At one extreme there may be just a single party involved, where the purchaser can provide the offset itself by undertaking appropriate management on other land that it owns (or acquires for this purpose). At the other, both of the purchaser and the provider may involve other parties in fulfilling the operational role of ensuring that the offset is delivered. The purchaser may not take any direct role in providing the offset beyond ensuring that someone is committed to supply it, and probably paying a regular or one-off financial contribution to support this.[95] Where any form of continuing commitment on the part of the purchaser exists (or is required, as discussed below), the operational element of this, such as checking relevant plans and monitoring progress, might be placed in the hands of a third party which has the expertise and capacity to carry out these tasks (for example, an environmental consultancy or a conservation non-governmental organisation (NGO)). Moreover, where the purchaser is a developer it may be appropriate to transfer any long-term burden in relation to the offset to those enjoying the benefit of the development once it is completed (perhaps under a conservation covenant as discussed in Chapter 5), thereby freeing the developer from future obligations.

In practice, further complications are likely to arise in the case of significant development projects. This is because of the many different ways in which such projects are structured and financed, so that there is not one single party who can be identified as 'the developer'. Even if control of the land where the development is taking place is regarded as being a key element, the position will vary depending on whether the land is being retained in the same hands throughout, acquired by the 'end-user' prior to development or bought specifically for the purpose of development and re-sale. The relationships between the various parties who can be seen as constituting 'the developer' will have to be

---

94 House of Commons Environmental Audit Committee, *Biodiversity Offsetting* (n. 88), paras 30–35.

95 The issues involved in providing long-term financial support are discussed in ICF Consulting, *Study on Specific Design Elements of Biodiversity Offsets* (n. 9) 66–74.

considered and provide for any obligations arising from an offsetting requirement.

Similarly, there may be several parties on the provider's side. If the provider is a landowner, then the actual management of the land might be placed in the hands of another party, such as a conservation body, under a lease or a contract to fulfil certain specific tasks. Alternatively, the provider might operate not on the basis of having its own land on which biodiversity projects can be undertaken but by agreeing with a range of landowners that their land is to be dedicated to the purpose. Moreover, there may be a role for brokers, who have the expertise and knowledge to match offset purchasers and providers in an efficient way that reduces transaction costs. This can be a purely intermediary role, or can be combined with some operational and monitoring responsibilities and can expand to evolve into a full bio-banking scheme, as discussed later.[96]

The second dimension of external involvement arises because offsets tend not to be wholly free-standing, but to operate within a wider context where the reason for the trade engages other parties. The offset may be required or encouraged as part of a permitting process, such as obtaining planning permission for development, so that the permitting authority must be satisfied that a satisfactory arrangement is in place; or there may be tax or other benefits at stake, again requiring that official recognition is gained, and maintained, in order to satisfy the authorities that the required action is being taken. As well as creating a scrutiny role for the public authority,[97] this may lead to additional parties being involved. For example, a regulatory authority may wish to pass on the long-term responsibility of monitoring the offset provision to another body which is better equipped to undertake this task, such as the relevant nature conservation body. This may be particularly appropriate in cases where the offset has consequences which the tax authorities have to recognise.

On the other hand, the overall pattern could be simplified if the regulatory authority takes on a stronger role and a direct relationship with the offset provider, rather than simply exercising a supervisory function over the arrangements between purchaser and provider. This would allow any reporting or enforcement functions to be simplified but does impose greater responsibility on the authority. The purchaser could either be allowed to drop out of the picture altogether, or retain responsibilities.

---

96 See section 4.7 below.

97 Under the US wetland mitigation scheme a conservation easement providing for the required compensatory measures must be officially approved no later than the harmful operations taking place; 33 CFR §332.7(a)(5), 40 CFR §230.97(a)(5).

Where there are continuing responsibilities there will be significant differences in practice depending on whether the purchaser is the party primarily responsible for ensuring that all is going well or merely a party to whom recourse can be had as a last resort in the event of the offset providers failing to fulfil their obligations.[98]

Even if an offset is not part of a formal regulatory scheme, some other assurance scheme may be in operation to enable the parties to demonstrate to other audiences that an offset deal is genuinely achieving its goals, for example, to show to investors or the public at large that those whose activities are causing biodiversity losses are compensating for this in accordance with commitments to corporate and social responsibility. The web of relationships can thus be complicated, with an offset fitting within a wider framework of commitments between parties, relating to specific tasks or broader objectives and for fixed periods or open-ended in time.

These relationships will all require legal obligations to bind them together. Where purely private parties are involved, these must be based on contract. The complexity here is that the long-term nature of the arrangements required to deliver offsets sits uncomfortably with the personal nature of contractual obligations. This means that such obligations normally come to an end in the event of one party dying, or are vulnerable (at best) in the event of insolvency or other changes affecting a party in the case of legal rather than natural persons. There may be ways round such difficulties and ways of establishing chains or networks of contracts to cover the most likely situations, but the fact that conscious recourse to such devices is necessary reveals the weakness in a purely contractual approach. In particular, where the obligation affects the use of land, securing this for the extended period required to deliver the gains cannot be achieved simply by a contract between the two parties, but must 'run with the land' in some way, so that future owners are also bound. Such enduring restrictions on land run counter to the general trend of legal policy which has been to free land from long-term encumbrances, and the position here is considered more fully in relation to conservation covenants in Chapter 5.

Where a public authority has a role, then contract may again be the chosen vehicle for creating legal obligations, but statutory intervention

---

[98] This would be analogous to the position of the owner of contaminated land in the United Kingdom, who bears liability for remediation if it is not possible to have recourse against a more appropriate 'responsible person' (such as the one who caused the contamination); Environmental Protection Act 1990, Pt IIA.

may allow for different forms of arrangement. This may be as simple as providing a means for contractual obligations to survive events which would otherwise bring them to an end, for example, providing that management agreements affecting land will continue to bind successive owners rather than terminating when the land is sold.[99] Alternatively, there may be specific statutory recognition of and provision for agreements with or obligations undertaken by relevant parties.[100] Beyond that, offset requirements may be embodied in regulatory mechanisms, for example, as conditions in the grant of planning permission for the development which gives rise to the biodiversity loss.[101] Where such a regulatory approach is followed, the nature of the obligations and the means of enforcement will be established by the regulatory framework in question, rather than depending on contractual means. Whatever the legal form of the arrangements, providing some binding commitment will be an essential part of developing a biodiversity offsetting as a credible means of conserving biodiversity.

4.5 ASSURANCE

If an offsetting scheme is to be widely accepted, people must have confidence in it, both the parties directly involved and the wider public. Those who wish, or are required, to acquire offsets to compensate for the damaging impacts of their activities must be assured that the offsets will provide the gains that are sought, especially if they are to retain any responsibility for ensuring that the offset is delivered. Those who are seeking to provide offsets, must have a way of showing that the gains they can offer are real and will be delivered. The wider public must have confidence that the scheme can deliver its promise of no net loss,[102] requiring assurance that losses and gains are being properly assessed and that the offsets will in fact produce what is promised.[103] The long

---

[99] See e.g. Countryside (Scotland) Act 1967, s. 49A.

[100] See e.g. Town and Country Planning (Scotland) Act 1997, s. 75.

[101] The pilot scheme in England discussed above operated on the basis of using the planning system as the vehicle for providing and securing offsets.

[102] In media discussions in England the label 'licence to trash' was used to express the fear that offset arrangements were simply a way of permitting valuable countryside to be destroyed in exchange for illusory compensatory measures.

[103] Analysis of decisions under the EU Habitats and Species Directive suggests that the relevant authorities are too willing to accept compensatory measures which are not really adequate; Ludwig Krämer, 'The European

time-scales involved pose particular challenges.[104] Key issues are establishing confidence that the offsets are genuinely offering at least a good prospect of achieving no net loss and that suitable enforcement methods will be available if things go wrong. Further supporting issues relate to how progress is to be monitored and to transparency at various levels.[105]

4.5.1 Valuation

At the initial stages, the valuation processes must be accepted as reliable. Naked self-interest provides an incentive for developers to minimise the scale of offset they require to provide and for offset providers to maximise what they say they can deliver. Therefore, some mechanism should be in place to supply assurances on the reliability and accuracy of the valuations on each side. Not only do the buyers of the offset need to be sure that what they are buying will in fact meet their needs, but regulatory bodies, such as planning authorities, may need confirmation that a satisfactory offset is being provided. More generally, those concerned with the fate of biodiversity need reassurance that the offset scheme is living up to its claims of being able to provide appropriate conservation measures.

The confidence required can be provided in several ways. Where offset requirements are part and parcel of official regulatory processes, such as where this is a mandatory requirement to obtain permission for a development to proceed, the task of carrying out valuations, or of ratifying them, might be put in the hands of the relevant public authority. This, though, places a burden on the authority and is likely to entail some duplication of effort, since the buyer and provider will have to carry out their own initial valuations in order to make their deal, before the approved valuation is undertaken, confirming or rejecting their calculations. The task of assessing biodiversity losses and gains could be entrusted to a conservation body, either a public body or one or more NGOs or other bodies certified for this task, placing this function where the ecological expertise already rests (and in the process perhaps creating through fees a new income stream to support conservation work). A

Commission's Opinions under Article 6(4) of the Habitats Directive' (2009) 21 *Journal of Environmental Law* 59.

[104] The need to secure long-term conservation benefits is one of the key issues explored by ICF Consulting, *Study on Specific Design Elements of Biodiversity Offsets* (n. 9).

[105] Gordon and others, 'Perverse incentives risk undermining biodiversity offset policies' (n. 36).

further option is to rely on individual assessors but subject to a certification scheme of some sort, based on either training and qualification for this specific task[106] or on some wider qualification, such as membership of a relevant group within a professional body. The need for approval or certification, whatever form it takes, will add to the transaction costs involved in an offsetting scheme, but seems inevitable if it is to gain widespread credibility.

4.5.2 Guarantees

For offsets to be acceptable, there must be confidence that the biodiversity gains being promised will actually be delivered; or rather, given the uncertainty discussed above over delivery in practice, that those responsible will do what they have promised to do with a view to providing the biodiversity gain. This focuses attention on the legal arrangements noted above. Legal obligations can be created through the normal rules of contract, through the rules of contract as adjusted to meet the special needs of the long-term agreements engaged here, or through statutory provisions. Crucial issues to determine are where responsibility is to lie for providing the offset and to what extent that responsibility can be transferred.

In the paradigm case of land that is about to be developed subject to a requirement to provide an offset in order to obtain permission for the development, the primary responsibility to ensure that suitable compensatory gains are delivered rests on the developer.[107] A restrictive scheme could state that the developer must itself provide the offset. In practice, the developer is unlikely to have the land or expertise to undertake this task, nor does that long-term responsibility fit with the commercial realities of the structure and funding of the development industry.[108] Developers may therefore wish to pass on to others the direct task of delivering the offset, but a separate question is whether they should be able to pass on the legal responsibility as well. One approach is to say

---

106 E.g. the procedure in New South Wales for BioBanking Assessor accreditation; NSW Government, *Biobanking Assessor Accreditation* (2015), available at www.environment.nsw.gov.au/biobanking/assessaccredit.htm. See ICF Consulting, *Study on Specific Design Elements of Biodiversity Offsets* (n. 9) 49–51.

107 Subject to the difficulties of identifying 'the developer' discussed in section 4.4 above.

108 Especially in view of the risks of development companies falling into insolvency or being wound up in a way which might extinguish any offset liabilities, as discussed below.

that the party which initially creates the need for an offset must continue to bear the responsibility for ensuring that it is provided. If this line is taken, then the developer cannot escape liability and is always the party against whom the planning authority will have recourse if the offset measures are not being taken as required. This provides a strong incentive for the developer to provide itself with as strong assurances as it can that those physically providing the offset can be relied on, and perhaps creates a new network of indemnity and insurance contracts to protect the developer against future liabilities. Indeed, some such arrangements may be required before the planning authority is willing to grant permission, as is the case when bonds to cover restoration costs are required as a prerequisite for permission for some mineral and waste sites.[109] Conservation covenants such as discussed in the next chapter may play a key role in establishing the requisite legal guarantees that the offset site will continue to provide biodiversity benefits into the future.

An alternative is to allow the developer to pass on legal responsibility, but again some assurance mechanisms are likely to be appropriate. The transfer may be simply to the successor as owner or occupier of the site; after all, since they are now getting the benefit of the development which has been allowed to proceed, it seems appropriate that they should also bear the connected burden. In such cases, stronger assurances are likely to be required to ensure that the responsibilities are taken seriously, especially as time goes by and the connection between the initial development and the potentially distant offset provision fades in memory.[110] An alternative to transferring responsibility to a party benefiting from the development is to transfer it to a body with the motivation and expertise to ensure that the offset is in fact provided. This may offer efficiencies, since such bodies, whether statutory or non-governmental, might well be taking an interest in the offset site in any case, and assuming the powers of the offset purchaser (presumably in exchange for a one-off or regular payment from the purchaser) might strengthen their position and simplify the relationship for all concerned. In such cases, the nature of the body taking over responsibility might reduce the need for formal assurance mechanisms.

Parallels exist with the transfer and surrender of environmental licences which may be permitted only when the relevant authority is

---

109 E.g. Council Directive 1999/31/EC of 26 April 1999 on the landfill of waste, [1999] OJ L182/1, art. 8(a)(iv).

110 Particularly for commercial buildings, the (re-)development cycles may be much shorter than the natural cycles involved in establishing habitat.

satisfied that suitable arrangements are in place to ensure future compliance with any licensing requirements or in the case of a surrender, that there are no existing or future liabilities. Thus in Scotland, the transfer of a permit may be refused if the Scottish Environment Protection Agency (SEPA) is not satisfied that the transferee will ensure compliance with the permit, or in the case of waste permits does not meet the statutory test of being a 'fit and proper person' to hold such a licence.[111] In relation to the surrender of licences (bringing to an end the permit holder's responsibilities), SEPA must be satisfied that 'all appropriate measures have been taken' to 'avoid pollution risk', 'return the site to a satisfactory state' and 'remove, control, contain or reduce any relevant hazardous substance in soil or groundwater so that the site … ceases to pose a significant risk to human health or the environment'.[112] Applying such approaches to biodiversity offsetting, the focus of any restrictions on when the original developer can bring to an end its responsibilities might thus be on the person to whom these responsibilities are being transferred or on the satisfactory completion of the steps required to provide the offset.

4.5.3 Insolvency, Winding-up and Restructuring

A constant thread running through the discussion of offsets is the long-term nature of the obligations on the various parties. This means that special attention has to be paid to what is to happen in the event of the original parties ceasing to exist or being restructured in a way that places the continuation of the obligation under threat. To the extent that public authorities are involved, these are subject to restructuring in various ways, but the statutory provisions achieving this usually[113] provide adequately for the transfer of rights and liabilities in a way that should cover the relevant obligations and duties, but this should be checked, especially since the potentially open-ended nature of some offset requirements may make them different from most of the other matters involved.

In relation to companies and associations, recourse will have to be had to the general rules on their merger, division, winding-up and insolvency. Such issues have already arisen in other areas of environmental law and have not always produced clear and satisfactory results. In the United

---

[111] Pollution Prevention and Control (Scotland) Regulations 2012, SSI 2012/360, reg. 47.

[112] *Ibid.* reg. 48.

[113] See n. 114 below.

Kingdom, problems have arisen over the liability of the companies that are successors to the nationalised Gas Board for land contaminated in the past,[114] the capacity of the receiver of an insolvent company to disclaim liabilities under a waste management licence,[115] and the responsibility of the liquidators of a mining company for the restoration of the land under statutory licences.[116] In such cases there is a clash between the principles of insolvency law, which may allow the abandonment of onerous property and liabilities, and those of environmental law seeking to make the polluter pay. In dealing with offsets, there is the further aspect that one may be dealing not with a specific liability that is apparent at the time that an insolvency is being worked through, but with what will happen if things go wrong at some unknown date in the future. The key point is that if offset arrangements are to endure, then attention must be paid to what is to happen in the event of disruptive events affecting the continuity of the parties in future.

4.5.4 Finances

It costs money to manage land so as to achieve biodiversity gains, and ensuring that secure finances are available for this is another significant aspect of providing assurance that an offset will be able to deliver the gains being sought over the long term. An initial issue is to determine where liability for the costs is to fall but in most cases one would expect the purchasers to be contributing at least some of the costs of providing the offsets they are benefiting from. Significant costs are likely to arise at the beginning of an offsetting project, both to meet the direct costs of undertaking work, such as planting or altering drainage so that a site can deliver biodiversity benefits, and to cover the time-lag before any income can be obtained from the site in its new use, such as from sustainable harvesting of new woodland. However, there are also the costs of maintenance which can also be substantial and potentially run on indefinitely.

---

114 *R (National Grid Gas plc (formerly Transco plc))* v. *Environment Agency* [2007] UKHL 30, [2007] 1 WLR 1780.

115 *Re Celtic Extraction Ltd (In Liquidation)* [2001] Ch. 475.

116 *Joint Liquidators of the Scottish Coal Co. Ltd* v. *Scottish Environmental Protection Agency* [2013] CSIH 108, 2014 SC 372. In this case there was further litigation about the activation of the bond provided to ensure that there were funds to pay the remediation costs; *East Ayrshire Council* v. *Zurich Insurers plc* [2014] CSOH 102.

The simplest mechanism is just to require a one-off payment by the purchaser at the beginning of the offset. This could either be just to meet the initial costs or could go beyond this to set up an endowment producing income to cover future costs, or a sinking fund, where both income and capital are used to meet future costs until the fund is wholly used up. Simply covering the initial costs inevitably leaves future maintenance to be paid for by other means, whilst a sinking fund will also last only so long into the future. Moreover, the fickleness of financial markets may mean that what seems like adequate provision today in future falls short of meeting the costs it is supposed to cover. In practice, trying to cover all eventualities may be unrealistic and counter-productive; after all, there are few, if any, situations where future costs are genuinely met into the indefinite future. All of these options, however, impose a sizeable capital burden on the purchaser, and the impact of this on the finances of development projects will have to be carefully considered if an offsetting scheme is to be feasible, especially since the timing is likely to mean that this funding is provided before any proceeds from the project can be realised.

A different approach, on its own or supplementing an up-front payment, is to provide for continuing contributions. These could come from the purchaser or it is more likely that these will be passed on to those benefiting for the time being from the project which necessitated the offset being provided. This may seem reasonable in the early years, but there may be difficulties if the project is not a financial success, especially as time goes on and the link between the development and the offset fades in the memory. There is a risk of such obligations being seen as obscure and unreasonable as liability for chancel repairs in England.[117] This is quite apart from the problems of setting up the formal legal mechanisms to enable the obligation to continue at a meaningful level decades into the future and to be transferred between successive owners, and where necessary divided if ownership is split, and to deal with situations such as insolvency or compulsory purchase.

---

[117] See the arguments in and response to *Aston Cantlow and Wilmcote with Billesley Parochial Church Council* v. *Wallbank* [2003] UKHL 37, [2004] 1 AC 546; Mark Reading, '"Practice what you preach": a guide to the development of chancel repair liability in the twenty-first century' [2008] *Conveyancer and Property Lawyer* 140; BBC News, 'Chancel repair liability: the ancient law that could hit house prices', 13 March 2014, available at www.bbc.co.uk/news/uk-england-26373756.

4.6 MONITORING AND ENFORCEMENT

It will be necessary to provide a suitable means of ensuring that whatever is required to deliver the biodiversity gains is in fact provided. Several stages are involved, requiring clarity on what obligations are placed on whom, monitoring to see whether those obligations are being fulfilled, legal standing for relevant parties to take action if they are not, and appropriate remedies when they do intervene. A further initial point is to emphasise the importance of clear drafting which will enable parties many years later to determine with certainty whether or not the terms of any agreement have been broken.[118] Nevertheless, whatever provisions are put in place for monitoring and enforcement, these will only be effective if there is a willingness to use them and to keep the long-term obligations alive. This is often where things fall down.[119]

4.6.1 Records

An essential starting point is that those concerned must know what obligations exist, whichever side of the duty they are on.[120] Where an offset agreement is connected to a regulatory scheme, for example, as a condition of obtaining a development permit, the official records of that process should preserve the details of the obligations. Similarly, if the agreement is embodied in a conservation covenant, the registration and related formalities which tend to accompany that mechanism are likely to provide enduring records.[121] More uncertain is the position where an offset agreement stands outside, or is not completely reflected in, an official record and registration scheme. The risk is that over the extended duration of the agreement its details, or even existence, are lost or forgotten as individuals come and go and parties change. It might, therefore be argued that all offset agreements should be registered or recorded in some way, but this is probably best addressed not by creating

---

118 See the discussion on contracts for PES in Chapter 3, section 3.5.

119 See e.g., the findings of limited enforcement activity in the English town and county planning system, despite the availability of a wide array of powers and mechanisms to this end: Adam Sheppard and others, *Planning Enforcement England: At the Crossroads, Project Report* (University of the West of England, 2014), available at http://eprints.uwe.ac.uk/24531/.

120 Cf. the importance of registries in carbon trading schemes: European Commission, 'Climate Action: Union Registry', available at http://ec.europa.eu/clima/policies/ets/registry/index_en.htm.

121 See Chapter 5, section 5.6.

a new central registry but by insisting that any offset arrangement which is to be given official recognition in any way does take a legal form which entails recording or registration, and that this is done in a way which includes sufficient details for future enforcement to be possible.

4.6.2 Standing

In view of the contractual nature of any offset agreement, enforcement will primarily lie in the hands of the parties. Any difficulties lie in ensuring that the contractual arrangements make suitable provision for the enduring nature of the obligations, allowing their benefit and burden to be passed on to successors. This can be achieved through assignment, by imposing obligations to create fresh binding legal ties when a new party takes on a relevant role, or as a matter of land law (through covenants or the like). Where the offset is embodied in a regulatory approval, whether as a condition of the permit or as a parallel agreement with the authority, then it will have standing to enforce its terms.

More interesting questions arise over whether other parties could and should have any standing to intervene. It would be against the fundamental contractual and private nature of an approach based on agreements to allow anyone other than a party to the agreement to have standing to enforce its terms. Yet since these agreements are entered to serve the public interest in conserving biodiversity, there is scope for arguing that ensuring compliance with the agreements is not just a private matter. If a specific party, such as a conservation body, is identified as external guarantor of a deal, then the rules on privity of contract might be circumvented by expressly including them as a party to the agreement, or its terms could be drafted so as to take advantage of the relevant rules of contract which allow rights to third parties.[122] Alternatively, a public conservation body could be given statutory standing to intervene to enforce private agreements when this is in the public interest but the relevant contractual party is not doing so.

Such blurring of the public-private divide will not occur where a public body has a direct role within a regulatory scheme, but the question arises of whether members of the public would have standing to intervene directly, or by challenging the failure of the regulatory body to take enforcement action. Such standing depends largely on the state of the domestic administrative law, where although the party intervening may

[122] In the United Kingdom, under the Contracts (Rights of Third Parties) Act 1999 in England and Wales or the *ius quaesitum tertio* in Scotland.

be expected to show some interest in the matter, this can be interpreted broadly[123] or narrowly.[124] Relevant non-governmental organisations can be given a privileged position to take action, as under the Aarhus Convention.[125] The most extreme position is to allow anybody to enforce an offset deal. This is the approach taken in New South Wales, where legislation provides that:

> Any person may bring proceedings in the Land and Environment Court for an order to remedy or restrain a breach of a biobanking agreement, whether or not any right of the person has been or may be infringed by or as a consequence of the breach.[126]

A compromise which enables the public to have a say whilst still protecting authorities from potentially unjustified litigation is to follow the model in the EU Environmental Liability Directive.[127] The Directive's liability provisions can be invoked only by the 'competent authority' designated for this purpose by the Member State, but members of the public and relevant organisations are able to call on it to take action. Such requests cannot be ignored and the authority must either begin to take action or, if it decides not to, provide reasons for this decision. This decision must then be susceptible to review in the courts or by whatever other independent review procedures the national legal system allows.[128]

4.6.3 Remedies

The right to take action is only meaningful if an appropriate remedy can be obtained. The English common law's view that damages should be the

---

[123] The ebbs and flows in the traditionally wide standing in the United States are discussed in Christopher Warshaw and Gregory E. Wannier, 'Business as usual: analyzing the development of environmental standing doctrine since 1976' (2011) 5 *Harvard Law and Policy Review* 289.

[124] As under the rules on title and interest to sue in Scotland until the relaxation announced by the Supreme Court in *AXA General Insurance Ltd, Petitioners* [2011] UKSC 46, 2012 SC (UKSC) 122 and confirmed in *Walton* v. *Scottish Ministers* [2012] UKSC 44, 2013 SC (UKSC) 67.

[125] Convention on Access to Information, Public Participation in Decision-Making and Access to Justice in Environmental Matters (Aarhus Convention), adopted on 25 June 1998, entered into force on 30 October 2001, Art. 9(1).

[126] Threated Species Conservation Act 1995, s. 127L(1).

[127] Directive 2004/35/CE of the European Parliament and of the Council of 21 April 2004 on environmental liability with regard to the prevention and remedying of environmental damage, [2004] OJ L143/56.

[128] *Ibid.* art. 12.

primary remedy for a breach of contract is recognised as not being appropriate in circumstances such as these where a financial sum will not make up for the true loss to biodiversity.[129] Injunctive remedies, preventing damaging activity and requiring the necessary positive steps to be taken, are therefore the likely remedies. The further question arises over the existence of any powers of direct intervention where any defaulting parties are unable or unwilling to take the steps required of them. It is not uncommon for environmental regimes to give a public authority the power to take direct action, where the responsible party is not acting or where urgency demands this, and then to recover the costs of this intervention.[130] This may already be part of a development permitting regime under which offset arrangements fall,[131] but otherwise might be considered as a means of ensuring that there is a route to practical intervention so that biodiversity interests do not suffer.

4.6.4 Rights of Entry

A subsidiary but important issue which arises relates to rights of entry. If they are to be able to carry out monitoring and enforcement roles, those charged with these responsibilities must be able to gain access to the sites in question. Only an on-site inspection will enable the state of the habitat to be ascertained and decisions to be taken on whether promised work has been undertaken properly, whilst effective monitoring may require setting up camera traps, water flow or quality monitors or other equipment in order to ascertain the state of the relevant environmental elements. Such activity is an intrusion on the property of the landowners and as such would normally require their consent. Such consent, however, may not be forthcoming, for a variety of reasons. There may be a desire to hide non-compliance; non-co-operation may be a side-effect of a wider breakdown of relations between the parties; there may be uncertainty over who can give consent;[132] or simple dilatoriness on the part of the owners may mean that express consent cannot be obtained when required. Several options to deal with that unsatisfactory situation are available.

---

[129] Law Commission, *Conservation Covenants: A Consultation Paper*, Law Com. No. 211 (2013) paras 6.42–6.45.

[130] See e.g. Nature Conservation (Scotland) Act 2004, s. 40; Pollution Prevention and Control (Scotland) Regulations 2012, SSI 2012/360, reg. 57.

[131] See e.g. Town and Country Planning (Scotland) Act 1997, s. 135.

[132] E.g. in the event of a messy succession or insolvency dispute.

The precise terms of any rights of entry are endlessly variable, for example, in relation to notice procedures or limitations on exactly how and how often they can be exercised, and the following discussion concentrates just on the key issue of there being some provision to ensure that those interested will be able to gain access to the land in question. The solution which is most in keeping with the private nature of the offset arrangement is to leave this to the parties. The parties could be free to set out in the terms of the offset deal whatever access arrangements, if any, they consider appropriate. This is likely to lead to varying results, but that is an inevitable consequence of relying on individually negotiated deals rather than imposing uniform standards. The risk of deeply unsatisfactory outcomes could be reduced by providing guidance or model terms so that the parties do at least have their minds directed to the issue and are provided with suggestions for appropriate terms to include. At the other extreme, there could be a legal provision inserting rights of entry as a mandatory element of any relevant agreement, either on an absolute basis or as the default position subject to express alteration by the parties. Setting the scope for such a provision might seem difficult (what sort of agreements are included?) and the intervention could be based on the legal form of the agreement rather than its specific role in providing an offset, for example, applying to all conservation covenants.[133] In practice, though, it is likely that offset arrangements will tie in with some other official processes, whether simple registration or having to be recognised in connection with obtaining development consent or financial benefits, and these could be used to set the boundary for the application of the statutory entry provisions. Having appropriate rights of entry could be one of the standards in judging whether what is proposed merits official recognition.

Further options look not to the agreement between the parties but give the right of entry to an external authority. Rather than providing a means for a party to insist on access, the existing powers of statutory bodies could be extended to allow them to intervene. This could be as part of their general functions, as in the case of official conservation bodies which often have wide powers of access,[134] or more specifically tied to the planning or other regime under which an offset is recognised, such as giving a planning authority the power to access land to ensure that the

---

133 See Chapter 5, section 5.6.

134 E.g. the powers enjoyed by those authorised by Scottish Natural Heritage under the Nature Conservation (Scotland) Act 2004, s. 44.

offset provision incorporated in the terms of the granting of approval for a project is being satisfactorily met.

4.7 BIO-BANKING

A development of offsetting is the establishment of a 'bio-bank'.[135] As so often in this area, the term can be used to cover a range of arrangements with significant variations, but the essence of a bio-bank is that projects which can provide biodiversity gains are recognised so that a purchaser wishing to offset losses can just buy a share in a project that is already under way rather than having to set up a new arrangement.[136] There are several advantages in this approach. From the perspective of the purchasers, the system is straightforward. The scale of the losses to be offset is measured according to the valuation rules that apply and then the appropriate number of credits is acquired from those operating an approved conservation scheme. The transaction costs (other than undertaking the valuation) are reduced, there are no enduring liabilities and official recognition is guaranteed in dealing with an approved bio-bank. The providers can also benefit since they can embark on conservation projects with the sale of credits generating an income stream to finance this work, rendering conservation a commercially viable form of land use. For them too, transaction costs are reduced and operating within an established framework reduces the uncertainty of ad hoc arrangements, but there is an element of risk if no buyer can be found for the credits generated. From the environmental perspective there are also advantages. Bio-banking schemes provide an incentive for larger-scale and longer-term projects than are likely to be inspired by ad hoc arrangements, and as such bring greater ecological benefits. They 'typically involve larger, more ecologically valuable parcels, and more rigorous scientific analysis, planning and implementation than permittee-responsible mitigation' (i.e.

---

[135] Sometimes known as a 'habitat bank', especially where the term 'bio-bank' has been used in relation to the storage of genetic material.

[136] The New South Wales BioBanking scheme is a leading example; for an introduction and further links see NSW Government, *BioBanking: A Market-Based Scheme* (2015), available at www.environment.nsw.gov.au/biobanking/ and for an early critique, Jane Scanlon, 'An appraisal of the NSW Biobanking Scheme to Promote the Goal of Sustainable Development in NSW' (2007) 4 *Macquarie Journal of International and Comparative Environmental Law* 71.

than individual offsets established by a developer).[137] The incentive for such larger projects is a tangible one in that the income available from selling credits renders conservation a commercially productive use of land.

As with other forms of offsetting, though, there are design challenges in establishing a scheme that will provide the certainty and security that all parties may desire, and practical difficulties in operation.[138] For the purchaser the key issue will be the availability of credits which will satisfy their needs in terms of scale and recognition. Offsetting is done for a reason, usually to meet regulatory requirements, therefore it is vital that the arrangements made enjoy the requisite official recognition. Other than checking that the credits are being acquired from an accredited bio-bank, the purchaser's main concern is likely to be over valuation. Here the same issues arise as under other offsetting arrangements, both in terms of scale of offset required and any restrictions of the sort of project which can be accepted, especially how far there are constraints requiring like-for-like replacement of the habitat being lost or other limitations on what is an acceptable compensatory provision. Once these hurdles have been overcome and the purchaser has paid for the credits being acquired, the purchaser essentially drops out of the picture, a position that may justify a premium in the payments for using a bio-bank rather than other forms of offsetting. The management of the offset site, the obligation to ensure that the site continues to serve its biodiversity purpose and any consequences and risk if the expected biodiversity benefits are not delivered, all pass to those running the offsetting project, who are in any event in the best position to deal with these. The bio-bank may take direct responsibility for operating the offset sites or act at one step removed. In the latter case, the bio-bank is also serving a role as broker, but will require a further layer of legal obligations to govern relations between the bio-bank and those 'on the ground'.

A lot therefore hangs on the recognition and calculation of the credits involved, both the number required and the number being provided. The calculation question raises nothing distinct from what arises in other forms of offsetting, but the recognition one does require special attention.

---

[137] US wetland mitigation rules, 33 CFR §332.3(b)(2), 40 CFR §230.93(b)(2). There is express mention of the benefits of banking schemes in 'consolidating projects ... and resources, providing practical planning and scientific expertise ... , reducing temporal losses of functions, and reducing uncertainty over project success': §290.93(a)(1).

[138] Salzman and Ruhl, 'Currencies and commodification of environmental law' (n. 61) 648–68.

Any bio-bank accreditation scheme must take account of both the capacity of any project actually to deliver biodiversity benefits in the future and how far the project is guaranteed into the future. Since bio-banking projects may be on a larger scale and planned on a more organised basis, rather than reacting to opportunities, there is scope for such issues to be satisfactorily addressed. One outstanding issue, though, may be that of additionality, especially if there is an over-supply of bio-bank credits. It is part of the essence of bio-banking that conservation projects begin and gather credits 'in the bank', ready for future sale, but there is the danger of a surplus of credits being built up. If a project has been going on for several years without its credits being used, does there come a point at which what it is providing should be seen as part of the current ecological baseline rather than as offering something additional? Taking that approach places a risk on the offset provider, reducing the incentive to become engaged in the process except where almost guaranteed a market. Such risk may well affect decisions as to what sort of biodiversity gains are produced, for example, favouring short-term and easily reversible gains over those which take longer to mature. On the other hand, the benchmark against which 'no net loss' of biodiversity must be judged is inevitably dynamic and it would seem wrong not to reflect any established improvements (however inspired) as well as further degradation in determining the starting point.

In terms of ensuring that the biodiversity gains are delivered, the challenges are the same as for other offsets, requiring appropriate long-term guarantees that the land will continue to be used in a way likely to deliver the benefits sought. The difference is that monitoring and enforcement will clearly be the responsibility not of the purchaser but of those with responsibility for running or overseeing the bio-bank. This may require appropriate contractual provisions between provider and bio-bank (if the bio-bank does not itself take on operational responsibility) or statutory provisions for a supervisory body to ensure that there are suitable mechanisms for monitoring and enforcing the obligations undertaken by the provider. The bio-bank can also serve as the holder of the funds, which can provide a long-term income for those managing the sites concerned. The bio-bank provides a secure mechanism for enduring commitments, insulated from the changes in resources and priorities of the parties over years to come.

Experience in the United States, where 'mitigation banking' is an established way of dealing with the need to take compensatory action when wetlands are affected, suggests that it is the long-term delivery and guarantee of conservation that can be the weak points. There is a difficulty here in determining the criteria for effective measures, since the terms of the relevant permits often fail to cover adequately all of the ecological aspects of a site, so that compliance with the formal requirements may not represent true compensation for what has been lost.[139]

4.8 CONCLUSION

Although the basic idea of a biodiversity offset is simple, creating a legally binding framework that will last for decades into the future and take account of the likely interests of the various parties is not straightforward.[140] Various substantial reports have produced different ways of highlighting and addressing the various issues to be considered. The BBOP Principles on Biodiversity Offsets call for a framework based on the following points: adherence to the mitigation hierarchy; limits to what can be offset; landscape context; no net loss; additional conservation outcomes; stakeholder participation; equity; long-term outcomes; transparency; science; and traditional knowledge.[141] A report for the EU identifies a longer list of issues to be addressed in the design of schemes: objectives; mitigation hierarchy; mandatory/voluntary approach; 'like-for-like' offsets or better; metrics; additionality; location; timing; institutional arrangements; regulatory bodies; instruments; land availability; standards and performance criteria; certification and accreditation; monitoring and reporting; compliance and enforcement; long-term management; and contingencies for failure.[142] A further report identifies the 'key legal considerations' as: identification of parties; legal nature of agreement; objective and scope of contract; rights and obligations; payment

[139] Kelly Chinners Reiss, Erica Hernandez and Mark T. Brown, 'Evaluation of permit success in wetland mitigation banking: a Florida case study' (2009) 29(3) *Wetlands* 907; Tammy Hill and others, 'Compensatory stream and wetland mitigation in North Carolina: an evaluation of regulatory success' (2013) 51 *Environmental Management* 1077.

[140] The ICF Consulting Services report suggests that key requirements for successful offsets schemes include: binding contractual agreements; long-term management plans; rights to manage the land for conservation purposes; obligations to secure such management and prevent changes in use; secure access to finance and safeguards against failure: ICF Consulting, *Study on Specific Design Elements of Biodiversity Offsets* (n. 9) 76.

[141] BBOP, *Biodiversity Offset Design Handbook* (n. 3).

[142] ICF and GHK, *Exploring Potential Demand For and Supply of Habitat Banking in the EU and Appropriate Design Elements for a Habitat Banking Scheme*, Final report submitted to Directorate-General Environment (2013).

structure; duration; risk allocation; monitoring and reporting; and remedies and dispute settlement.[143] Further studies in specific contexts draw attention to similar issues.[144] However the issues are divided up, such analyses capture the potential and the challenges in developing offsetting schemes that will deliver real biodiversity benefits.[145]

From a legal perspective, it is the long-term nature of the commitment in most cases that renders legal guarantees both more necessary and more problematic. Changes over time in the priorities, capacities, motivations and even identities of the key parties, as well as fading memories of what was agreed, all point towards the desirability of clear and enforceable obligations if there is to be full confidence that the biodiversity gains promised are going to be created and maintained for decades to come. Yet putting these in place may be a hurdle that gets in the way of any progress towards providing compensatory action.

The alternative of simply trusting that everything will work out well may seem naïve, but in practice might be the better option in some circumstances. This is especially the case where arrangements can be made with trusted and responsible bodies which are not likely to disappear in the foreseeable future and whose commitment to conservation objectives is bolstered by external constraints, such as their narrow charitable purposes. Dealing with such bodies can allow very simple and effective arrangements to be made, such as developers making a one-off financial payment to support conservation projects that are already in progress, and that deliver a tangible biodiversity gain.[146] Albeit lacking legal guarantees, such arrangements have every prospect of longevity and avoid significant transaction costs.

The experience of carbon markets may have useful lessons. A study by Arup and Zhang[147] traces developments within some of the main trading schemes and notes that developments have been in the direction of tightening the controls on the markets, emphasising that an offsetting approach does not replace regulatory intervention but shifts its focus

---

143 Cascade Consulting (with eftec), *Assessing the Potential for Payments for Ecosystems Market Mechanisms* (Welsh Government, 2014) 76.

144 See e.g. Kormos, 'Great apes and biodiversity offset projects in Africa' (n. 55).

145 IUCN, *Biodiversity Offsets Technical Study Paper* (n. 26).

146 As in the schemes arranged by Scottish Borders Council; see www.scotland.gov.uk/Resource/0042/00426294.pdf; and discussion in Chapter 7, section 7.4.

147 Christopher Arup and Hao Zhang, 'Lessons from regulating carbon offset markets' (2015) 4 *Transnational Environmental Law* 69.

from the practical activity to the market framework that allows for offsets.[148] Even in the much simpler area of carbon emissions, it has been found to be 'technically challenging to gauge whether many offsets are genuine reductions',[149] whilst assessing additionality has been a particular problem. The various changes made within the markets have been towards a tightening rather than a relaxation of the rules, limiting the proportion of work towards meeting targets that can be contributed by offset arrangements and restricting what can be accepted as an offset.[150] Their overall conclusions are pessimistic:

> Offset regulation has weak spots, if not fatal flaws … Offsets therefore remain unreliable as instruments to mitigate climate change … Still, our studies have found positive signs … Properly regulated, it is possible that offsets may still play a useful role in climate mitigation.[151]

Doubtless others might present a more encouraging picture, but there are clearly risks in placing too much reliance on offsetting.

As noted earlier, though, accepting the existence and operation of an offsetting scheme does not commit one to seeing this as the primary mechanism for achieving conservation.[152] Nor does it entail accepting that every feature of the ecosystem is 'up for grabs' and vulnerable to damage so long as an offset can be offered. There is scope for excluding some areas altogether, for setting standards that restrict the forms of compensation that are acceptable or require the achievement of more than just 'no net loss', and for ensuring that the offset does not result in a temporary loss to the ecosystem. There are significant challenges in creating a legal framework that can provide firm guarantees for the interests of all the parties concerned over the lengthy and potentially indefinite duration of an offsetting deal, but solutions can be found. The role to be played by offsetting will determine the nature of the arrangements established and for a variety of reasons, that role may be limited. Offsets may not be appropriate for providing the main route to achieve conservation but may have a useful role in respect of residual losses, providing a means of ensuring that biodiversity is not completely ignored

---

[148] *Ibid.* 96.
[149] *Ibid.* 100.
[150] *Ibid.* 97.
[151] *Ibid.* 100.
[152] The operation principles adopted by the EU's Working Group on No Net Loss of Ecosystems and their Services state that: 'Biodiversity offsets cannot be a "primary" policy tool to be implemented in all cases'; see http://ec.europa.eu/environment/nature/biodiversity/nnl/pdf/NNL_Operational_Principles.pdf.

when other factors justify undertaking projects that inevitably have some adverse consequences. If there are practical gains to be achieved in comparison with the present position, that might provide a strong argument for an acceptance of some offsetting arrangements, despite a recognition of conceptual and operational weaknesses.

5. Conservation covenants

Controlling how specific areas of land can be used in order that they will benefit biodiversity is a mainstay of current conservation policy. In most countries there are various designations of land, such as National Parks, Nature Reserves, or Special Areas of Conservation, which carry with them more or less severe limitations on how the sites can be used in order to secure habitat within which biodiversity can flourish. The same idea lies behind conservation covenants (also known as conservation easements and by other names). The vital difference is that whereas the designations rest ultimately upon an exercise of regulatory power by public authorities, conservation covenants are voluntary and operate within the sphere of private law. This distinction is not absolute; some of the controls within officially designated sites may be a matter of agreement rather than legal regulation, while conservation covenants may be agreed with statutory bodies and may operate within a framework of statutory provisions that make alterations to the general private law rules. Moreover, there is the risk of a supposedly voluntary agreement to enter a covenant being de facto required in exchange for regulatory approval or public benefits of some other kind.[1] Nevertheless, by being rooted in private law and the choices of private parties, such covenants mark a significant change from the more conventional approach. They can provide a mechanism for securing biodiversity offsets, as discussed in the previous chapter, as well as allowing for individual initiative in determining what features of land are to be given enduring protection.

Creating the potential for significant use of conservation covenants usually entails a departure from narrower statutory schemes which provide for agreements regulating land use by enabling agreements to be made by a wider range of parties, for a wider range of purposes and for longer periods. They also represent a departure from the standard rules of

---

1 Jessica Owley Lippmann, 'Exacted conservation easements: the hard case of endangered species protection' (2004) 19 *Journal of Environmental Law and Litigation* 293; Jessica Owley Lippmann, 'The emergence of exacted conservation easements' (2005–06) 84 *Nebraska Law Review* 1043; Jessica Owley, 'The enforceability of exacted conservation easements' (2011) 36 *Vermont Law Review* 261.

private law by overcoming limitations on the use of enduring restrictions on the use of land. The interaction with the existing legal background, especially the rules of land law, is important and means that this mechanism will vary between jurisdictions depending on the wider private law. Even where the covenants are self-contained new creations as opposed to adaptations of existing devices, the wider context will have played a significant part in how they were designed and will affect the terminology. The name used varies between jurisdictions, matching the wider usage in land law: covenants (as proposed for England and used in New Zealand), easements (United States), burdens (Scotland) or servitudes or some other term. Since in practice at least some aspects of these mechanisms are the product of fairly recent legislative intervention, it is the content of the statutory provisions, not the label, which matters, but the label will nevertheless be significant in shaping expectations. Thus, the fact that the proposals in England are for the establishment of conservation *covenants*, not easements, carries with it an expectation that these covenants can be revised or overridden through recourse to some judicial procedure, as is the case for other covenants between landowners,[2] but not for easements. These devices do not automatically and necessarily take on all the features of the mechanism whose name they may share, but the label will undoubtedly affect how they are perceived. In this book, 'covenants' is used as the generic term because it does not carry as many specific connotations as the more precise terms used in some jurisdictions.

The key features of conservation covenants were identified by the Law Commission for England and Wales which has recently been examining the potential of introducing conservation covenants there:

> A conservation covenant is an agreement made between a landowner and a conservation body which ensures the conservation of natural or heritage features on the land. It is a private and voluntary arrangement made in the public interest, which continues to be effective even after the land changes hands.[3]

This highlights several points. First, there is the dual nature of the arrangement as a private agreement serving a public purpose, which raises questions over whether it can be left wholly in the hands of the parties or should be subject to a degree of public scrutiny. Secondly, there

---

2 Law of Property Act 1925, s. 84.

3 Law Commission, *Conservation Covenants*, Law Com. No. 349 (2014) para. 1.1.

is the essential element that the agreement has enduring effect, 'running with the land' and binding successive owners, not just the one who entered it. Thirdly, it shows that consideration must be given to further issues, notably the nature of the party with whom the landowner can agree and the purpose for which the agreement can be made. Other factors which are important in deciding the shape of any scheme for covenants include the duration of the agreement, the extent to which it can be varied or discharged and the enforcement mechanisms. These issues are inter-related, since the broader the range of parties and purposes, the greater the need for some scrutiny to ensure that covenants are not used inappropriately.

Underlying such concern is the fact that in creating enduring restrictions on the use of land, conservation covenants are going against the tide of centuries of development in land law, at least in the common law world. Evolving from an era of entails and other burdens that attempted to dictate the future of land indefinitely into the future, English law has been concerned to reduce the proliferation of enduring restrictions on land which will have an impact on land as an asset which can be freely used and transferred by the current holder.[4] This concern was exported to the Commonwealth and US jurisdictions which share the common law as a root. Moreover, until the comparatively recent development of land registration (whether of titles or of obligations affecting the land)[5] there was a further concern over the difficulties of being able to keep track of

---

[4] Such development reflects the view which has been dominant in recent centuries, treating land primarily as a commercial asset. That view is far from uncontested, e.g. by the arguments for a stewardship approach to land-holding; William Lucy and Catherine Mitchell, 'Replacing private property: the case for stewardship' (1996) 55 *Cambridge Law Journal* 566.

[5] In Australia, registered titles based on the Torrens system have been in place since the mid-nineteenth century (Real Property Act 1858 (South Australia)). In England and Wales, registration of titles and of land charges was introduced by legislation in 1925 (Land Registration Act 1925, Land Charges Act 1925) but despite recent efforts over 15 per cent of the land is not yet covered by registered titles; HM Land Registry, *Annual Report and Accounts 2013/14* (HM Land Registry, 2014) 43. In Scotland, registration of deeds in the Register of Sasines dates back to the early seventeenth century, with registered titles introduced by the Land Registration (Scotland) Act 1979 and presently including 58 per cent of titles covering 26 per cent of the area of Scotland but with a recent commitment to complete registration by 2024 (Registers of Scotland, *Completion of the Land Register: Public Consultation* (Registers of Scotland, 2014) 2.

what rights exist and who can enforce them.[6] As a consequence, the law was willing to recognise as binding on successive owners only those restrictions which benefit neighbouring land. Easements or covenants 'in gross', for the benefit of more distant parties rather than nearby landowners, were not allowed; this restrictive view is generally not a feature of the civil law jurisdictions of continental Europe and those they have inspired.[7] The broad approach in England was confirmed as recently as 2011 when the Law Commission in a general review of easements and covenants rejected introducing obligations which were not for the benefit of neighbouring land;[8] subsequent exploration of conservation covenants has been seen as being a discrete exception in response to the growing interest in private (as opposed to state-run) conservation action.[9]

Land law in other jurisdictions developed differently, such as in Scotland where elements of the feudal system continued to provide the basic structure of the law until major reforms in the first decade of this century. One feature of this was that the rights of the feudal superior enabled the creation and enforcement of enduring restrictions and conditions on the use of land, regardless of whether neighbouring land was held. This device was used to provide long-term controls over land use in a range of circumstances, including the development of the elegant New Town in Edinburgh in the late eighteenth century and the more functional 'new towns' created after the Second World War.[10] When the future of such feudal burdens was considered as part of a comprehensive programme of reform, started in the 1990s and led by the Scottish Law Commission, the initial view was that there should be no special provision to enforce burdens imposed for reasons other than the amenity or service of other nearby land.[11] The result of the consultation process,

---

6 Several jurisdictions still do not have a comprehensive system of registration, and in others, as mentioned in the preceding note, the legal framework is in place but it is proving a slow task to ensure that all titles and obligations are in fact registered.

7 For the position in relation to conservation servitudes in South Africa, where a mixed legal tradition produces results similar to the common law, see Alexander Paterson, 'Tax incentives: valuable tools for biodiversity conservation in South Africa' (2005) 122 *South African Law Review* 182, 187–9.

8 Law Commission, *Making Land Work: Easements, Covenants and Profits à Prendre*, Law Com. No. 327 (2011) para. 2.24.

9 Law Commission, *Conservation Covenants: A Consultation Paper*, Law Com. No. 211 (2013) paras 1.9–1.12.

10 E.g. Cumbernauld and Glenrothes.

11 Scottish Law Commission, *Property Law and the Abolition of the Feudal System*, Scot Law Com. DP No. 93 (1991) paras 3.30–3.32.

however, was to persuade the Commission that such burdens should be recognised. The prime examples given and the weight of representations related to burdens affecting the future use, alteration and maintenance of historic buildings, without which local authorities and trusts might be unwilling to provide financial support for restoration work, but the potential for this mechanism to be used for wider environmental purposes was also recognised.[12] Accordingly, the legislation that followed permitted existing feudal burdens of this type to be converted into burdens that could be enforced under the new law[13] and enables certain new burdens to be created.[14]

The general antipathy to enduring restrictions on the use of land has not prevented the recognition of obligations which bind not just the original parties but also successive owners of the land affected, but these have often been very constrained. In the United Kingdom, many individual statutory provisions have allowed for management agreements and other arrangements between public authorities and landowners which do run with the land and can be enforced even though the authority does not hold any neighbouring land which is directly benefited. Where such statutory arrangements differ from conservation covenants is that the parties who can enter them and the purposes for which they can be made are much more constrained, and although they can be permanent they are often in effect only for a fairly short period rather than providing long-term controls.[15] Some of these provisions are narrow, such as for the provision and maintenance of cattle grids on roads,[16] whereas others are very broad, such as the power of Scottish Natural Heritage to enter agreements 'to secure the doing of, whatever in the opinion of the parties to the agreement may be necessary to secure the conservation and enhancement or to foster the understanding and enjoyment of the natural heritage of Scotland'.[17] The bodies who can make such agreements

---

[12] Scottish Law Commission, *Report on Abolition of the Feudal System*, Scot Law Com. No. 168 (1999) paras 4.52–4.61 and *Report on Real Burdens*, Scot Law Com. No. 181 (2000) paras 9.1–9.25.

[13] Abolition of Feudal Tenure etc. (Scotland) Act 2000, ss. 26–32.

[14] Title Conditions (Scotland) Act 2003, ss. 38–48.

[15] For a survey of such agreements in England and Wales see Law Commission, *Conservation Covenants* (n. 9) 124–44.

[16] Highways Act 1980, s. 87.

[17] Countryside (Scotland) Act 1967, s. 49A.

include Ministers,[18] the Forestry Commission,[19] the statutory conservation bodies[20] and the National Trust.[21] This power is also available to and widely used by planning authorities who can supplement the conditions imposed when granting planning permission with further agreements which can create significant and enduring obligations under which the developer contributes to resolving any adverse effects of the project being allowed to proceed (which can include ad hoc biodiversity offsets).[22]

The pressure to accept broader provisions for conservation covenants comes from several sources. One strand, notably in the United States, has been the absence of more direct regulatory control on land use. The general view of the limited role of the state in controlling the use of private property, and concerns over the point at which any restriction becomes a 'taking' prompting an entitlement to compensation,[23] mean that the level of direct regulation is much less than in Europe. Moreover, the pressure for regulation has been less since the pattern of land ownership is also very different with substantial areas, including many of the most notable areas for biodiversity and natural beauty in public ownership, as opposed to being in private hands. If substantial restrictions are going to be put in place on private lands, this has to be by agreement and to meet this need legislation to recognise conservation easements has been passed across the nation.[24] In many cases, this legislation follows the Uniform Conservation Easements Act produced by

---

18 See e.g. Ancient Monuments and Archaeological Areas Act 1979, s. 17.

19 Forestry Act 1967, s. 5.

20 See e.g., Wildlife and Countryside Act 1981, ss. 27A–33; Nature Conservation (Scotland) Act 2004, ss. 3–22, 39–49; Natural Environment and Rural Communities Act 2006, s. 7 (England); National Parks and Access to the Countryside Act 1949, s. 16 (Wales); Conservation (Natural Habitats, etc.) Regulations 1994, SI 1994/2716, reg. 16 (Scotland); Conservation of Habitats and Species Regulations 2010, SI 2010/490, reg. 16(4) (England and Wales).

21 National Trust Act 1937, s. 8. See also the provisions in New Zealand for Open Space Covenants with the Queen Elizabeth II National Trust (Queen Elizabeth the Second National Trust Act 1977, s. 22); Debra L. Donahue, 'Law and practice of open space covenants' (2003) 7 *New Zealand Journal of Environmental Law* 119.

22 The contribution can be purely in financial terms; Town and Country Planning Act 1990, s. 106; Town and Country Planning (Scotland) Act 1997, s. 75.

23 See Chapter 2, section 2.9.

24 R.H. Levin, *A Guided Tour to the Conservation Easement Statutes* (Land Trust Alliance, 2010).

the National Conference of Commissioners on Uniform State Laws.[25] The spread of such easements has been greatly assisted by the presence of tax breaks which in essence treat entering an easement with a charitable body as a gift to charity of part of the value of the land, which can be set off against income from other sources.[26] This encourages the use of easements but does distort the detailed arrangements, with meeting the requirements for tax relief potentially being more significant than the needs of conservation, as discussed later.

Elsewhere, the role of covenants will be to supplement the direct regulatory provisions operated by public authorities, enabling greater private initiative. This may arise from landowners wishing to ensure protection for things that they value even when they no longer control the land themselves, either once it is in the hands of future generations or as they realise some of the commercial value of the property by selling it. Equally, it may arise from conservation bodies keen to augment the protection provided through statutory schemes. In either case the parties are free to agree for themselves which features are to be protected, so that habitats and species other than those given top priority within the official conservation measures can be given protection and support, whilst in an age of austerity private funding can be used to augment what the public authorities can achieve. Conservation covenants may also be a means of giving effect to the long-term commitments which are at the heart of biodiversity offsetting arrangements.

The potential usefulness of conservation covenants has been recognised,[27] with value being seen in moving beyond long-term agreements in very specific circumstances to a more open system. Yet the fear of creating a proliferation of restrictions on land which hampers its utility

---

25 National Conference of Commissioners on Uniform State Laws, Uniform Conservation Easement Act (1981, last revised 2007), available at www.uniformlaws.org/shared/docs/conservation_easement/ucea_final_81%20with%2007amends.pdf.

26 Zachary Bray, 'Reconciling development and natural beauty: the promise and dilemma of conservation easements' (2010) 34 *Harvard Environmental Law Review* 119; Daniel I. Halperin, 'Incentives for conservation easements: the charitable deduction or a better way?' (2011) 74 *Law and Contemporary Problems* 29. The tax position also raises a more pressing need for a financial valuation of easements; Julie Ann Gustanski and John B. Wright, 'Exploring net benefit maximisation: conservation easements and the public-private interface' (2011) 74 *Law and Contemporary Problems* 109.

27 E.g. in England, see Green Balance, *The Potential of Conservation Covenants: A Report by Green Balance for the National Trust* (National Trust, 2008).

and tradability means that there remains a desire to keep this mechanism within bounds. Covenants are thus constrained and a number of key issues arise in designing the framework which is to be used, questions which are inter-related since the balance struck on one aspect may increase or decrease the need for controls elsewhere. In essence, the choice is between creating a very loosely defined mechanism but then introducing controls on its operation so as to ensure that the worst dangers can be avoided, or restricting the ways in which covenants can be used, for example, by limiting the purposes and parties which are eligible, but then taking a more hands-off approach for those that pass this eligibility test. Examples of both approaches can be seen below. The other tension that runs through all of these issues is that between conservation covenants' legal form as a private agreement and the recognition that there is a public interest in the goals they are serving.

5.1 PURPOSES

Where the background land law does not permit enduring restrictions on land for non-neighbours, the scope of any specific exception for conservation purposes must be determined. Practice has been to state the eligible purposes in fairly broad terms, covering both the natural and cultural heritage. Thus, the Scottish legislation states that conservation burdens must be:

> for the purpose of preserving, or protecting, for the benefit of the public:
>
> (a) the architectural or historical characteristics of any land; or
> (b) any other special characteristics of any land (including, without prejudice to the generality of this paragraph, a special characteristic derived from the flora, fauna or general appearance of the land).[28]

Noteworthy is the requirement for the purpose to be for the public benefit, which is designed to emphasise that this departure from the standard rules of land law is to exist as a limited exemption to serve public, not private, purposes. A difficulty is, of course, that some public conservation will also benefit private interests. If some of the woodland surrounding a house is made subject to a protective covenant, that will

---

[28] Title Conditions (Scotland) Act 2003, s. 38(2). Similar burdens can also be created for the purpose of reducing greenhouse gas emissions: Title Conditions (Scotland) Act 2003, s. 46A, added by Climate Change (Scotland) Act 2009, s. 68.

benefit biodiversity but probably also the amenity (and value) of the house;[29] if areas of wetland are conserved to maintain water quality, this benefits not only the aquatic ecosystem but also anyone who abstracts the water downstream for commercial or public utility purposes.[30] In the Scottish definition, so long as the public benefit is clear then a private by-product will not be a problem, but the public focus must be retained.

A less restrictive approach is taken in the Uniform Conservation Easements Act in the United States which permits obligations:

> the purposes of which include retaining or protecting natural, scenic, or open-space values of real property, assuring its availability for agricultural, forest, recreational, or open-space use, protecting natural resources, maintaining or enhancing air or water quality, or preserving the historical, architectural, archaeological, or cultural aspects of real property.[31]

Since the purpose of the easement need only include the conservation goals, this opens the door to easements where there is considerable private benefit also being achieved. Indeed, one criticism of the practice in the United States has been that the support given (both through the legal acceptance of easements and the financial benefits that accompany this as a result of the tax system) has been either misdirected or disproportionate in view of the public benefits achieved.[32] The US provision also extends to agricultural and forest uses, not just those concerned with conserving the natural or cultural heritage.

The lesson here, as noted above, is that if there are concerns about the potential proliferation and misuse of covenants, these can be largely avoided by providing only a narrow scope for the mechanisms, but at the expense of restricting its usefulness and potentially excluding desirable applications. On the other hand, a looser approach may create opportunities for inappropriate uses and lead to calls for some form of direct scrutiny of actual applications to ensure that these are excluded.

---

29 In Western Australia those entering nature conservation covenants are expressly told by the Parks and Wildlife Department to note that 'in some peri-urban situations, the covenant may in fact result in an increased valuation due to a desirability for lifestyle blocks of conservation value'; see Government of Western Australia, *Nature Conservation Covenant Program FAQs* (2013), available at www.dpaw.wa.gov.au/management/off-reserve-conservation/nature-conservation-covenant-program/142-faqs#rates.

30 Law Commission, *Conservation Covenants* (n. 3) paras 3.20–3.23.

31 Uniform Conservation Easements Act, s. 1(1).

32 Halperin, 'Incentives for conservation easements' (n. 26) 41–3.

5.2 PARTIES

It would be possible to enable landowners to enter covenants with any other party, but this has not been the approach taken in either the United States or the United Kingdom. By limiting the parties, a number of purposes are served. It limits the proliferation of such burdens, it constrains their scope to those within the *vires* of the bodies listed, and it should be a protection against inappropriate burdens since the specified bodies are hardly likely to accept terms which are thought unworkable or to place inappropriate obligations on either party. Moreover, such bodies will be expected to have the resources, expertise and long-term commitment to maintain adequate records and to take their responsibilities seriously in terms of enforcement, not least because of the extent to which they can be called to account, legally (and politically in some cases), by the public and/or their members and beneficiaries.

The Scottish model is to limit the covenant holders to the Scottish Ministers or one of the conservation bodies designated for this purpose by the Ministers.[33] The conservation burden is extinguished if the holder of it ceases to be a designated conservation body[34] and burdens can be assigned only to another conservation body or the Scottish Ministers.[35] Moreover, if a burden is created by someone else (for example, by a private landowner as opposed to by a conservation body over land it is disposing of), it can take effect only with the consent of the conservation body concerned or the Ministers.[36] The bodies designated include all Scottish local authorities, the National Trust for Scotland, a number of heritage and building preservation trusts and, in relation to the natural environment, Scottish Natural Heritage, Plantlife, the John Muir Trust, the Royal Society for the Protection of Birds, the Scottish Wildlife Trust and the Woodland Trust.[37] In the United States, the Uniform Conservation Easements Act permits easements to be held by governmental and charitable bodies, the latter status being particularly significant in relation to the tax breaks which have been so influential in practice.

---

[33] Title Conditions (Scotland) Act 2003, s. 38. This approach is also proposed for England; Law Commission, *Conservation Covenants* (n. 3) ch. 4.

[34] Title Conditions (Scotland) Act 2003, s. 42.

[35] *Ibid.* s. 39.

[36] *Ibid.* s. 38(2).

[37] Title Conditions (Scotland) Act 2003 (Conservation Bodies) Order 2003, SSI 2003/453 (as amended). Only bodies whose objects or functions are or include the purposes noted above can be designated; Title Conditions (Scotland) Act 2003, s. 38(5).

By limiting those who can hold a conservation burden, some aspects which might otherwise have to be regulated are essentially taken on trust, but in a way that does open up the process to a degree of scrutiny by at least a portion of the public. Any system that depends on maintaining a list can, however, become administratively burdensome, risks becoming out-of-date and can be seen as limiting the flexibility being sought by moving only a short distance away from the current position where enduring restrictions can be agreed only under specific schemes involving a few specific public bodies. Yet this may be a simpler and less onerous approach than allowing greater freedom to enter and deal with covenants but having to impose more direct supervision and regulation.

Two further drawbacks of this listing approach are also worth noting. The first is that to the extent that Ministers and local authorities are involved, the fact that these have many responsibilities may mean that there is an internal conflict between maintaining and enforcing a conservation burden and pursuing other legitimate policy goals, such as permitting economic development on land currently dedicated to conservation and protected by a covenant. An answer to this is to point to the political accountability of such public bodies and the fact that they are expected to represent the public interest, balancing competing objectives in all sorts of contexts. Nevertheless, this may give rise to concern that the supposed guarantee created by a conservation covenant is in fact far from secure, a point related to the issue discussed shortly of how much flexibility there is to be in relation to the modification and discharge of the covenant. The second drawback is that the approved list excludes any commercial body, even those such as utilities which may have a strong 'public service' element in their activities. Thus, for example, the proposed scheme for conservation covenants in England will not be available to water companies there (which are in the private sector) even though they might be a useful mechanism for securing long-term protection of catchment areas to reduce pesticide use, etc. so as to maintain the quality of drinking water and reduce treatment costs (as discussed in Chapter 3). Against a background of accepting only limited exceptions to the general rule against enduring restrictions, however, it may not be inappropriate to exclude such arrangements from the current proposals in view of their different purpose and the different questions which they raise.

5.3 DURATION

The issue of how long a covenant is to last is integrally connected to whether and how it can be modified or brought to an end, yet since slightly different considerations arise at each stage it is worth considering them separately. Across all these issues there is a fundamental tension between permanence and flexibility. On the one hand, there is a desire to guarantee that the conservation aims of any covenant will be secured into the future, so that those aims can actually be realised; indeed in many cases, such as establishing new woodland, long-term provision is essential if anything meaningful is to be achieved at all. On the other hand, if the land is subject to permanent restrictions it may be harder to persuade landowners to enter arrangements which bind not just them but their successors a long way into the future, whilst changing circumstances and priorities may result in provisions that seem sensible today being undesirable of even counter-productive in the future.

The fact that covenants are meant to last is one of the features which distinguishes them from the many short-term management agreements and other schemes which are a common feature of conservation. For example, the agri-environment schemes developed as a means of giving effect to the EU's agriculture policy involve many options for farmers to commit themselves to particular forms of land management or specific activities, but for a limited period.[38] This serves the interests both of the farmers, who generally are reluctant to commit to very long-term arrangements in view of the volatility of agricultural markets and policy, and of governments, which prefer not to be tied in to long-term funding arrangements which commit specific spending years into the future. It is to move beyond such short-term schemes that covenants come into play, both to secure the permanence of any biodiversity gain, especially important in the context of offsets where the gain being secured is intended to compensate for biodiversity losses elsewhere, and to allow for the delivery of longer-term gains, such as allowing any new or restored habitat to develop into maturity, which in the case of woodland or areas such as boreal or alpine sites where plants grow slowly can take decades if not centuries.

---

[38] E.g. the various schemes under the current Scottish Rural Development Programme, which includes support for many agricultural activities with environmental benefits; see Scottish Government, *Rural Development Programme* (2015), available at www.gov.scot/Topics/farmingrural/SRDP/SRDP20142020 Schemes.

Such considerations point towards the desirability of covenants being made in perpetuity, guaranteeing for ever that the land will be used in the ways prescribed. The presence of a perpetual restriction gives certainty to all parties, ensures that in any offsetting scheme the permanent losses to biodiversity are balanced by permanent gains and makes sure that there is time for habitat to mature so as to achieve the maximum benefit. Nevertheless, there is a danger that seeking the security of permanent restrictions may actually not be in the wider interests. At least three aspects of this can be identified. First, at the initial stage, experience is that many landowners are reluctant to enter agreements which will tie the hands of their successors.[39] For some landowners who know or suspect that their heirs (or other successors as a result of sale or any other transfer) have very different views of nature from their own, tying the hands of successors may be a reason for entering a covenant or other long-term agreement in order to protect some features which cannot be safely left at the mercy of future owners. For many others, though, they do not want to dictate to their children and grandchildren or to tie them in to enduring commitments which may no longer make practical or commercial success in years to come. Leaving some flexibility to cope with the uncertainties of the future is seen as desirable and thus a reason for avoiding any permanent covenant.

A second consideration is that in view of the uncertainty of the future, any permanent restriction may become pointless. Measures intended to conserve a particular species or ecological community may become worthless if the arrival of disease or invasive species means that the intended beneficiaries are no longer there to be conserved. Coastal erosion, landslides, climate change affecting temperature and weather patterns and human development on neighbouring sites (or distant ones in the case of projects such as dams which can affect hydrological conditions long distances away) can all mean that the land governed by a covenant is no longer suitable for the purpose it was intended to serve.[40] With this in mind, burdening the use of land with permanent restrictions which cannot achieve useful results is absurd.

Changing policy and priorities present a third reason for questioning the desirability of truly permanent restrictions. Changes in our scientific understanding may mean that what we now think as being helpful for biodiversity is in fact of no benefit, or even harmful, as shown by our

---

39 This view is confirmed through conversations with practitioners advising farming clients on whether to enter various agri-environmental schemes.

40 Jessica Owley, 'Conservation easements at the climate change crossroads' (2011) 74 *Law and Contemporary Problems* 199.

developing understanding that in some circumstances apparently very destructive fires are not an environmental disaster but an essential part of the lifecycle of many ecosystems.[41] Society's needs may also change. A particular site set aside for conservation may in future be essential for some major infrastructure project of overwhelming importance. A general societal shift or the pressing needs of some crisis may lead to consensus that past conservation objectives should be overridden.

All these reasons suggest that despite the benefits of covenants that are binding in perpetuity, providing for that position without any chance of escape might not be desirable. This is the outcome of the debate in England, where the proposal is for the default position to be that covenants are perpetual, but subject to the parties agreeing to a shorter duration and various means for the terms to be modified or discharged.[42] A further limitation flows from the basic rules of land law[43] in that tenants will not be able to create covenants that last for longer than the remaining term of their lease. In the United States, a different pattern is followed because of the interaction between conservation easements and tax law. In order to gain the benefit of charitable status, the easement has to be treated as a gift of rights in the land and therefore must be in perpetuity.[44] As will be seen below, this has consequences for the extent to which modification of the easement may be possible.

5.4 MODIFICATION

The extent to which it is possible for the terms of a covenant to be modified or discharged (discussed in the next section) is another important element in how far it goes to guarantee the future of conservation on any piece of land. Preventing all change risks the disadvantages noted above, whereas too free an approach to modification undermines the long-term security which covenants are supposed to give. It follows from the nature of covenants as a private agreement, that the parties should be

---

41 See e.g., the work of the Association for Fire Ecology, www.fireecology.org.

42 Law Commission, *Conservation Covenants* (n. 3) paras 5.36–5.44.

43 As summed up in the maxim *nemo dat quod non habet*, nobody can give away rights that they do not own.

44 Nancy A. McLaughlin, 'Conservation easements: perpetuity and beyond' (2007) 34 *Ecology Law Quarterly* 673; Clemens Muller-Landau, 'Legislating against perpetuity: the limits of the legislative branch's powers to modify or terminate conservation easements' (2009) 29 *Journal of Land Use and Environmental Law* 281.

free to amend them as they wish, but to the extent that covenants have been given special recognition because of their public benefits (and can attract public funding, either directly or through tax breaks), it can be argued that there should be some public oversight of the process. The law also has to deal with the situation where amendment is desirable but one party will not cooperate in effecting this.

An initial consideration is in relation to the parties. It is a fundamental feature of a covenant that the parties bound by the covenant will change over time, with the obligation running with the land so as to bind successive owners. Where the owner's interests are split, whether by division of the land or of the owner's rights, for example, by creating a lease, the rules of land law for other more established covenants protecting the interests of neighbours should provide a clear indication of how the situation is to be handled, although some express provision might be desirable to avoid uncertainty. Where the benefits of a covenant are restricted to certain parties, there must be some provision for adjustment of the parties, but keeping within the confines of the restrictions imposed. In the case of public bodies holding covenants, it is far from unlikely that the structure of the bodies will change over the decades that a covenant will endure, but there are likely to be general statutory succession provisions that govern the position. Where the class of holders is extended to include other designated bodies, such as non-governmental organisations (NGOs), the integrity of this approach is secured by provisions such as those in Scotland where conservation burdens can be transferred between designated bodies or the Scottish Ministers, but not to any other party.[45]

The private law character of a covenant suggests that the parties should be free to amend its terms as they wish and as discussed below the focus of debate shifts to the issue of whether there should be any limits on this. A different issue arises in the absence of an agreement between the parties to modify a covenant. The general law of contract would suggest that no change can be made, but it has been recognised that this position may be undesirable in relation to long-term restrictions on land. The absence of agreement may stand in the way of landowners using their land in ways which make sense from several perspectives, yet the covenant holder may be withholding consent unreasonably or in extreme cases not be contactable or able to execute the formal legal steps

[45] Title Conditions (Scotland) Act 2003, s. 39.

necessary.[46] To deal with such situations the law can provide a means to by-pass the unco-operative party and gain approval through reference to some other body authorised to grant consent. Thus, in Scotland, applications to modify a conservation burden can be referred to the Lands Tribunal[47] which is given a wide power to grant any application if it is reasonable to do so, but is required to pay attention to certain factors:[48]

(a) any change in circumstances since the title condition was created (including, without prejudice to that generality, any change in the character of the benefited property, of the burdened property or of the neighbourhood of the properties);
(b) the extent to which the condition:
  (i) confers benefit on the benefited property; or
  (ii) where there is no benefited property, confers benefit on the public;
(c) the extent to which the condition impedes enjoyment of the burdened property;
(d) if the condition is an obligation to do something, how:
  (i) practicable; or
  (ii) costly,
  it is to comply with the condition;
(e) the length of time which has elapsed since the condition was created;
(f) the purpose of the title condition;
(g) whether in relation to the burdened property there is the consent, or deemed consent, of a planning authority, or the consent of some other regulatory authority, for a use which the condition prevents;
(h) whether the owner of the burdened property is willing to pay compensation;
(i) if the application is [related to a change of ownership], the purpose for which the land is being acquired by the person proposing to register the conveyance; and
(j) any other factor which the Lands Tribunal consider to be material.

Such provisions deal with the danger of a covenant becoming ossified without any means of changing obligations which are no longer a good idea. In doing so they involve a degree of scrutiny to ensure that there genuinely is a need for change, including consideration of the public interest. Such consideration is lacking where the parties themselves use their freedom of action to agree a modification. This has been seen as a danger where the holder of the covenant is a multi-purpose body, such as

---

[46] Avoiding this position is one of the arguments for limiting covenant holders to a constantly up-dated list of designated bodies, as in Scotland.

[47] Title Conditions (Scotland) Act 2003, ss. 90, 98, 100.

[48] *Ibid.* s. 100.

Ministers or a local authority. Whereas organisations devoted to conservation will be limited by their legal purposes[49] to action with environmental benefits, these broader bodies have to serve a wider variety of interests. The fear is that although such bodies may enter covenants in good faith, intending to provide enduring protection of biodiversity, in future years the pressures created by their many other responsibilities may lead to a change of priorities and thus a willingness to abandon a covenant. Especially in the case of a covenant protecting a biodiversity offset, this may destroy a carefully balanced trade-off between environment and economy, and experience in related areas suggests that reassessments of the position will almost certainly be one-way traffic towards less protection for environmental concerns. This phenomenon is shown in the UK planning system where approval for 'bad neighbour' developments of many sorts[50] are granted subject to conditions limiting operating hours in order to protect the interest of neighbours, but are followed some time later by further applications to relax those conditions, at which stage there is often a reassessment resulting in less protection for neighbours.[51] For this reason it has been argued that since conservation covenants are given special status because of their role in the public interest, there should be some scrutiny to reflect the public interest when the parties propose modification, but this has generally been rejected.[52] A compromise, might be not to tie the hands of the authority but to require some further procedural step (such as public consultation) before a covenant can be modified, so that this cannot occur as a minor and overlooked element in a major project, but requires a conscious act which will focus attention on the issue, both within the authority and in the public eye.

A rather different picture can emerge where achieving charitable status has played a major role in shaping the obligations and the arrangements that embody them, as with most conservation easements in the United States (and there can be further considerations arising from the interplay

---

49 Embedded in statutory rules or the express purposes of a company, trust, charity or association.

50 Ranging from 'fast-food' shops to airports.

51 At a grander scale the same process can be seen in the erosion during the nineteenth and early twentieth centuries of the land and other rights embodied in the supposedly permanent agreements between US authorities and the native Americans; Walter Echo-Hawk, *In the Courts of the Conqueror: The 10 Worst Indian Law Cases Ever Decided* (Fulcrum Publishing, 2010) ch. 7 in relation to the Medicine Lodge Treaty.

52 Law Commission, *Conservation Covenants* (n. 3) paras 7.21–7.36.

with the Treasury requirements governing the tax breaks obtained).[53] Here, the rules of charity law require the agreement to be in perpetuity, precluding any possibility of modification in future, regardless of how desirable this may be to all concerned and how sensible it would be in serving the interests of conservation. This raises the question of what can be done when the terms of the easement are not merely no longer desirable, but actually impossible to fulfil, for example, if the species the easement is expressly designed to conserve is no longer present on the land concerned. Here, it has been argued that some amendment is possible, but subject to the limits of the cy-pres doctrine, a general doctrine of charity law which establishes that in the event of the specified charitable purposes becoming impossible, assets and obligations can be applied for alterative purposes, but only ones which fall within the particular underlying intentions of the party who originally established the arrangement.[54] External approval from a court or other official body is usually required. This may allow for some modification, but to a limited extent and in the case of significantly changed circumstances there may be no way of furthering a fairly narrowly expressed intention on the land concerned. This illustrates that the benefits of linking conservation easements to charity law come at a price.

5.5 TERMINATION OF A COVENANT

The considerations just discussed apply even more strongly in relation to the ending of a covenant. Again, the private nature of the agreement points towards the parties having a free hand in bring a covenant to an end but there are arguments that, especially in relation to offset arrangements, there should be some scrutiny of such a move to ensure that the public interest is being maintained, or at least procedural provisions to ensure that termination is not done inadvertently. These arguments have not generally been accepted, although the English proposals are that unilateral discharge by the holder of a covenant should be possible only where this is expressly provided for in its terms.[55]

---

[53] Jeff Pidot, 'Conservation easement reform: as Maine goes should the nation follow?' (2011) 74 *Law and Contemporary Problems* 1, 17–20.

[54] Nancy A. McLaughlin, 'Rethinking the perpetual nature of conservation easements' (2005) 29 *Harvard Environmental Law Review* 421; Jessica Owley, 'Changing property in a changing world: a call for the end of perpetual conservation easements' (2011) 30 *Stanford Environmental Law Journal* 121.

[55] Law Commission, *Conservation Covenants* (n. 3) paras 7.2–7.20.

Questions also arise over the application of other more general legal rules. In the first place there is the application, or not, of a doctrine of general land law, that of merger.[56] The standard position in common law based systems is that where the holder of an obligation affecting land comes to acquire full title to the land, the separate obligation disappears; the owners of the land cannot owe an obligation to themselves. Yet with conservation covenants, there are arguments that the covenant should survive, especially where the holder is a multi-purpose body since, as noted above, the commitment to conservation may be at risk from the other priorities it must serve. This is another aspect of the law where express provision is desirable, and the proposals in England are for the covenants to survive such merger, a conclusion swayed by the pragmatic point that it was undesirable for a covenant to be ended by accident when a relevant body acquired the land on a temporary basis as one stage of a bigger transaction involving the land ending up in other hands, a scenario that is not unlikely as local authorities and conservation bodies implement long-term arrangements for the land.[57]

There may be some overlap here with the question of whether a conservation covenant can be overridden by other statutory provisions, notably those allowing land to be compulsorily acquired by the state for public purposes (sometimes referred to as the exercise of powers of 'eminent domain'). Each jurisdiction has its own rules on such acquisition and again there is a tension between the public interest in seeking to preserve the conservation benefits which led to the covenant being created and granted legal recognition,[58] and the desire to fulfil the other public goals for which the land is being acquired, such as major infrastructure projects. The general position is that compulsory acquisition overrides all previous rights and obligations affecting the land and this seems the most appropriate response here as well, as so often hoping, but with no guarantee, that the public authority involved will give weight to conservation as well as its other responsibilities in determining when and how to exercise its powers.[59]

---

[56] Nancy A. McLaughlin, 'Conservation easements and the doctrine of merger' (2011) 74 *Law and Contemporary Problems* 279.

[57] Law Commission, *Conservation Covenants* (n. 3) paras 5.99–5.125.

[58] Nancy A. McLaughlin, 'Condemning conservation easements: protecting the public interest and investment in conservation' (2007–08) 41 *UC Davis Law Review* 1897.

[59] Law Commission, *Conservation Covenants* (n. 3) paras 7.101–7.115.

5.6 ENFORCEMENT

Two main areas of discussion have emerged in relation to the enforcement of conservation covenants. The first is ensuring that the covenant holder has effective forms of recourse when the terms of the covenant are being breached by the occupiers of the land; the second is whether there should be any form of intervention where a covenant holder does not act to enforce a covenant which is being breached. On the first issue, guidance can be taken from the wider rules of land law, whereas the second raises directly the rather ambiguous nature of covenants as private in form but to some extent public in function. A few other matters deserve attention first.

One preliminary issue to mention in passing is the importance for enforcement of all parties being aware of their rights and obligations. In many jurisdictions this will be achieved by a system of registration which applies to all substantive rights and interest in land,[60] supported in some cases by the limitation of the range of covenant holders to bodies which can be expected to be equipped to keep careful track of their own rights. To those from jurisdictions where registration is the norm this seems uncontroversial and any uncertainty is over the further issue of whether there should be any central register of covenants to enable the public to gain an overall picture of what land is subject to what restrictions (and where public funds may be supporting this).[61] In other places, though, even the idea of registration may be debatable, with some opposition in the United States based on fears regarding privacy and attracting unwelcome visitors.[62]

Regular monitoring and reporting is another means by which covenants can be kept alive and not forgotten. For example in Maine, laws have been introduced requiring the holders of conservation easements to

---

[60] See nn. 5 and 6 above.

[61] Law Commission, *Conservation Covenants* (n. 3) paras 5.71–5.87; Amy Wilson Morris and Adena R. Rissman, 'Public access to information on private land conservation: tracking conservation easements' (2009) *Wisconsin Law Review* 1237.

[62] James L. Olmsted, 'The invisible forest: databases and the end of the clandestine conservation of natural lands' (2011) 74 *Law and Contemporary Problems* 51. Such fears have to be seen in the light of the very different backgrounds of the rights of landowners in different jurisdictions, e.g. in the United Kingdom there are extensive statutory rights of public access to private land, especially in Scotland; Countryside and Rights of Way Act 2000, Pt 1; Marine and Coastal Access Act 2009, Pt 9; Land Reform (Scotland) Act 2003, Pt 1.

register the easement annually and to monitor the sites concerned at least every three years, retaining written copies of monitoring reports.[63] Such provisions ensure that there continues to be an active relationship between the easement holder and the landowners bound by the obligations, reducing the risk of slipping in to breaches, or of breaches remaining undiscovered.

A further preliminary point is to note that in order to assist enforcement it is desirable for a covenant holder to have access to the land affected so that the state of the land and of any required or prohibited activities can be ascertained. This can be expressly provided in the agreement creating a covenant, but it has been argued that such powers of entry and inspection should be provided by statute to ensure that covenant holders are in a position to monitor compliance.[64]

In terms of remedies, a preliminary issue is whether this is a matter purely for the civil law or whether there should be criminal sanctions as well. Using the criminal law to protect areas of land designated for their conservation value is part of the traditional regulatory approach and this could be extended to land covered by a conservation covenant. The covenant would thus take the place of the statutory designation, with actions in breach of its terms being a criminal offence. This is the case in some places, but the general view has been that this runs counter to the notion of covenants as private law mechanisms based on the agreement of the parties.[65]

Turning to the civil law, the more general rules of land law will provide the basis for the remedies when obligations such as a conservation covenant are breached, and the wider rules of procedure will determine how effective and accessible these will be in practice. Court procedures which are slow and expensive may deter formal action for all but the most serious of breaches. The availability of interim relief, ordering the cessation of activities while the full position is ascertained, and of an injunction or interdict when activities in breach of a covenant are anticipated but have not yet occurred, may also be important. The starting point is likely to be injunctive relief since the whole point is to prevent, and if necessary restore, the land to fulfil its conservation purposes. This underlying objective may limit the circumstances in which a court considers that damages should replace an injunction or specific

---

[63] Pidot, 'Conservation easement reform' (n. 53) 8–13.

[64] Law Commission, *Conservation Covenants* (n. 3) paras 6.14–6.24. See the discussion of rights of entry in relation to offset agreements at Chapter 4, section 4.6.

[65] Law Commission, *Conservation Covenants* (n. 3) paras 6.21–6.22.

performance as a remedy,[66] but there are situations where a covenant holder may deserve to be compensated. Interestingly, for England and Wales, the Law Commission has recommended that this should be one of the now very rare circumstances in which the court should have the power to award exemplary damages for flagrant or irreparable breaches of the landowner's obligations.[67]

Since the covenant may also include obligations on the part of the covenant holder, the landowner too should have a range of suitable remedies available. Here, though, the nature of the obligations, for example, to provide financial support, material assistance or guidance, is likely to more amenable to the standard financial remedies for breach of contract. Where the covenant holder is a public authority or a charity, there may be further means of recourse available if it is considered that they are not fulfilling their responsibilities, such as judicial review, complaints to an ombudsman or remedies within the system for the oversight and regulation of charities.

It is obvious that the parties to a covenant should be able to enforce its terms, and the nature of covenants as a private law mechanism suggests that these are the only parties who should have this power. Yet the quasi-public nature of conservation covenants raises the question of whether others should have the power to intervene if the holder is unable or unwilling to enforce the covenant. One extreme response is to open up enforcement to the world at large. This is the position in New South Wales where any person can take court action to enforce a 'bio-banking' agreement, 'whether or not any right of the person has been or may be infringed by or as a consequence of the breach'.[68]

More generally the discussion has centred on whether some public authority should have the power to intervene. In several States in the United States, this power has been expressly conferred on the Attorney General, whilst it is argued that the role of this officer as guardian of charitable trusts also confers power to intervene, given the charitable nature of the covenant holders in that context.[69] The issue has been discussed in the context of the proposals for England, but the idea of the Attorney General or some other body having the power to intervene was rejected. This conclusion was based on the view that such powers would detract from the essentially private and voluntary nature of covenants,

---

[66] In England, this is a general discretion enjoyed by the courts; Senior Courts Act 1981, s. 50; County Courts Act 1984, s. 38.

[67] Law Commission, *Conservation Covenants* (n. 3) paras 6.104–6.118.

[68] Threatened Species Conservation Act 1995, s. 127L.

[69] Pidot, 'Conservation easement reform' (n. 53) 13–16.

and on the practical difficulties of ensuring effective external engagement.[70] Such reasoning must, however, be considered in the context of the proposals to limit covenant holders to a restricted range of designated bodies, who can be expected to retain the desire and capacity to enforce the covenants they hold, reducing the likelihood of significant breaches passing without a response. In this, as in so many aspects of the design of a framework for covenants, the various pieces must be looked at together and against the wider legal background in the jurisdiction concerned in order to determine whether and in what form specific legislative intervention is required.

5.7 DRAFTING AND PARALLEL AGREEMENTS

All the above considerations emphasise the need to take care in the specific drafting of any covenant. Whatever the legislative framework created, express provisions in a covenant can deal with many of the issues which might otherwise give rise to problems or uncertainty in relation to duration, modification, discharge, enforcement and what is to happen in changing circumstances. The long-lasting nature of the obligations also requires attention to be paid to the specific terms that are being agreed. The more specific the substantive terms, the clearer the position is for all concerned in relation to what exactly is required of the landowner and whether a breach has occurred. Yet very specific terms are more likely to become outdated or to be overridden by external events.

Both the objectives and the particular obligations are affected by this issue.[71] In terms of objectives, linking a covenant with conserving a particular endangered species provides a strong and clear focus for all concerned, but may become pointless if disease, climate change or other factors mean that the land affected no longer offers a suitable habitat for that species. From this perspective a better option is to phrase the goals more generally, such as referring to conserving native woodland rather than a particular species, although drastic local changes might render even such broad terms inappropriate. An alternative is to refer to dynamic standards, such as compliance with the management standards as revised

---

[70] Law Commission, *Conservation Covenants* (n. 3) paras 4.101–4.111.

[71] Duncan M. Greene, 'Dynamic conservation easements: facing the problem of perpetuity in land conservation' (2004–05) 28 *Seattle University Law Review* 883; Jesse J. Richardson, Jr, 'Conservation easements and adaptive management' (2010) 3 *Sea Grant Law and Policy Journal* 31; Owley, 'Conservation easements at the climate change crossroads' (n. 40) 224–7.

from time to time by some official or respected conservation body. Similarly, there are short-term advantages in specifying particular management or other activities which must, or must not, take place but these may well become inappropriate as circumstances, and our knowledge and experience, evolve. A compromise is to draft the covenant in broad terms but to have more specific, shorter-term agreements running alongside it. These can either be separate, legally binding contracts running for a few years or looser arrangements, such as requiring in the covenant that management is carried out in accordance with subsequent guidance or agreements between the parties.[72]

5.8 CONCLUSION

The idea of enduring obligations affecting how land is used is far from new, but moving beyond more confined contexts to make these broadly available to a wide range of parties to serve broad conservation purposes does represent a significant change. Conservation covenants (or easements or burdens) offer an attractive means of providing more and more varied protection for land than is possible under the official public designations of protected areas and of involving a much wider range of parties, capitalising on the willingness of landowners and conservation groups to participate in the conservation enterprise. Although building on a private law mechanism, they may well require adjustments in order to overcome rules designed to protect land becoming unduly burdened by the 'dead hand' of past owners and the existence of a willingness to make such adjustments is a reflection of the fact that such covenants are not exclusively a private concern but also touch on public interests. Striking the balance between the private and public nature of such arrangements is a thread that runs through all the decisions to be taken in designing the legal framework for such agreements, a framework that will be affected by the wider legal background affecting rights and interests in land in each jurisdiction. The position can be further affected by the interplay with other legal rules, such as those on charity, since in some contexts an attraction of dedicating land in this way is the charitable nature of such action and its attendant tax breaks.

---

[72] A further interesting proposal, combining ideas discussed in this book, is for tradable conservation covenants, which allow a fixed conservation objective to be served on different parcels of land over time as conditions change; W. William Weeks, 'A tradable conservation easement for vulnerable conservation objectives' (2011) 74 *Law and Contemporary Problems* 229.

The use and popularity of covenants will vary according to the general context.[73] Where there are strong direct controls on land use and a far-reaching regulatory system seeking to secure conservation goals on private land, then covenants will play a complementary role to the more direct measures. Where such measures are weaker, then the importance of covenants may increase. Governments can promote covenants by creating tax breaks or other incentives for their use, although the rules on those may become a dominant influence on practice, whilst covenants can also be a vehicle for other mechanisms, such as payment for ecosystem services and biodiversity offsets. Covenants therefore have the potential to be an efficient and effective means of assisting conservation, but they can also be an expensive (through generous tax treatment) and cumbersome device, protecting features under no real threat and becoming an obstruction to flexible management of the land in decades to come.

The widespread use of covenants is too recent to gauge their true significance, bearing in mind that it is their enduring nature which is so fundamental to their operation. Anecodotal evidence in the United States suggests that the real test is yet to come, as conservation easements come under pressure from a new generation of landowners who may not share the same conservation interests as their predecessors who created the easements, and who have not seen the direct financial benefits of such arrangements. At the same time the original enthusiasm and memory of some easement holding bodies may be dimming, so that some easements may fade away rather than serving their long-term purposes. It will take a long time before it is clear whether covenants are a useful, vibrant and effective tool for conservation or become an obstructive anachronism and a hidden trap for property owners.[74]

---

[73] On the pros and cons of conservation easements in different situations in the United States, see Anna Vinson, 'Re-allocating the conservation landscape: conservation easements and regulation working in concert' (2006–07) 18 *Fordham Environmental Law Review* 273.

[74] As is the case with liability for chancel repairs in England, as noted in Chapter 4, section 4.5.4.

6. Taxation

Since the collection of taxes is pre-eminently a governmental function, a discussion of taxation measures may seem out of place in a book about the 'privatisation' of nature conservation. Nevertheless, it belongs here because it shares a crucial feature with the mechanisms discussed in the previous chapters, namely that it places decisions on what is going to be conserved and how, not in the hands of a state body, but in those of individuals. There is a governmental role in relation to the rules creating and regulating a market in ecosystem services or biodiversity offsets, but the practical results rest on the cumulative effect of many individual decisions, not on the imposition of regulatory controls or direct intervention. Similarly, tax systems may endeavour to shape decisions in pursuit of conservation goals, but they leave it to individuals to determine how far they respond to the incentives and disincentives created.

The operation of the tax system can have an impact on biodiversity in many ways, encouraging or dampening activities that cause harm, or promoting or obstructing those that are beneficial. Some of these consequences are deliberate, but often the impact is an unintended (and perhaps even un-noticed) by-product of tax rules devised with quite different considerations in mind. In the UK context, the role of tax breaks in promoting unsuitable forestry plantations in the 1980s was a rare example of the issue attracting attention but there are many other rules which do or might play a significant role, from the exemptions for heritage property in relation to inheritance tax to the special treatment of conservation bodies which qualify as charities.[1] Elsewhere, tax considerations also play a significant role, for example in the United States, the availability of tax breaks where the granting of a conservation easement qualifies as a charitable donation has played a major role in the use and form of this mechanism there.[2] Even issues apparently far-removed from nature conservation can have an indirect effect, as illustrated by the

---

1 See discussion of these examples later in this chapter.
2 See Chapter 5.

potential for changes to fuel duties to promote demand for biofuels, and in turn lead to changes in land use with consequences for biodiversity.[3]

Writing about taxation is challenging because there are so many variables involved, in terms of what is being taxed, the basis on which that is calculated or valued, who is liable to pay and the basis for any differential treatment designed to promote (or avoid unintended negative consequences for) biodiversity. The handling of the revenues once collected must also be considered. This is not the place for a comprehensive review of taxation systems, and since each jurisdiction's tax regime is comprised of a complex and inter-locking mix of very detailed provisions, the examples are primarily chosen from one system, the United Kingdom, rather than presenting isolated instances from wider experience which might not be properly appreciated without their wider context. What follows gives a brief and selective outline of certain key features before concentrating on some issues of particular significance in relation to taxes and biodiversity. First, there is some consideration of the basis on which tax is charged, then of the ways in which special treatment can be provided, before more particular points are considered.[4]

6.1 WHAT IS BEING TAXED?

Deciding what is going to be taxed is obviously a basic step in designing any tax regime and the potential impact on biodiversity will vary according to the subject of taxation. Within each category, the precise definition used in setting the boundary of the tax, and any exceptions or qualifications, will in turn affect the extent to which the tax will influence behaviour so as to promote or threaten biodiversity.

The taxes with the most obvious link to biodiversity are perhaps land taxes, since taxes based on the value of land are likely to encourage commercially productive uses, generating an income to meet the charge, rather than devoting them to nature. Land taxes are not a universal feature of national systems. In the United Kingdom, for example, they do not play a major part in the taxation system except for the rating system

---

[3] Alison Campbell and Nathalie Doswald, *The Impacts of Biofuel Production on Biodiversity: A Review of the Current Literature* (UNEP-WCMC, 2009).

[4] The extent to which any tax measures may be seen as 'state aid' under EU law, and thus lawful only if certain criteria are met and approval gained, is not explored here, but may be a consideration in practice; Treaty on the Functioning of the European Union (2007), [2012] OJ C326/47, art. 107.

for local authorities,[5] and even here, as discussed below, exemptions mean that most of the countryside is not directly affected.[6] Nevertheless, there is potential for land taxes to be a more important element of the tax system, as in some other countries, which would create a potentially stronger driver for or against different forms of land use with consequential impacts on habitat. The presence of transaction taxes affecting transfers of land, such as the stamp duty land tax,[7] may also be relevant, affecting the willingness of owners to make arrangements for the land that include a formal transfer of relevant property rights. If inheritance is a taxable event, then this might increase the pressure for land to be made commercially productive so as to meet the inevitable tax liability, and as a by-product encourage the fragmentation of extensive holdings as portions are sold off to satisfy the tax demand.

One important feature of land taxes (and other property taxes) is that the value of the land (and hence the liability to tax) will depend on the valuation method used. Aside from the wider issue of whether and how legal and financial systems can evolve to take into account non-commercial considerations, as is being tackled by the various attempts to assess and incorporate the value of ecosystem services, even the current valuation models may give habitat measures an indirect financial weight. The extent to which valuation takes account of potential as well as actual use provides a link between financial considerations and the impact of any restrictions (as a matter of regulation or operating in private law) which ease or constrain future changes in land use. Land where permission for profitable development is likely to be granted is more valuable than land where such development is not likely to be possible. Therefore the imposition or lifting of conservation measures standing in the way of

---

[5] The emphasis here is on the business rates levied on non-residential property rather than the council tax levied on domestic premises. Different but broadly similar systems operate for England and Wales, Scotland and Northern Ireland; UK Government, 'Business Rates', www.gov.uk/introduction-to-business-rates/overview.

[6] The removal of the exemption for shooting estates in Scotland is controversial, with differing opinions on the impact on biodiversity. See the papers leading up to and the consideration of the Land Reform (Scotland) Act 2016 (ss. 74–6); Scottish Government, Land Reform (Scotland) Bill (2015), available at www.gov.scot/Topics/Environment/land-reform/LandReformBill and Scottish Parliament, Land Reform (Scotland) Bill (2015), available at www.scottish.parliament.uk/parliamentarybusiness/Bills/90675.aspx.

[7] Finance Act 2003, Pt 4; and now in Scotland, the land and buildings transaction tax (Land and Buildings Transaction Tax (Scotland) Act 2013).

such a change in use may have a significant financial effect, as shown by the willingness to litigate over the designation of land under conservation laws.[8]

Where income is taxed,[9] the treatment of particular sources may significantly affect their attractiveness, for example, if certain payments for ecosystem services were not counted as taxable income, this might be a major incentive to seek such payments. By aggregating receipts and expenses from various sources to determine net taxable income, potential is created for encouraging certain activities by allowing the expenditure on these to be set off against income from other sources. The fine details of the tax rules may affect the ways in which particular forms of business are undertaken, such as by delimiting the categories of expenses which can be considered in calculating profits. The rules can also be drafted so as to encourage taxpayers to divert their own resources in particular ways across a broader canvas, such as by allowing donations to charities to count as a credit against income from any source, without the need for any link between the charitable purposes and the source of income.[10] If capital gains are taxed,[11] then the approach to valuation and any exemptions will be significant, for example, discounting any potential value that will not actually be realisable because of conservation restrictions.

In the case of a purchase or service tax, such as value added tax,[12] again there is some potential for influence on behaviour through the details of what can and cannot be taken into account at different stages. The greater impact, though, is likely to be on the definitions of and exceptions to what is taxable, affecting both the liability of those who have to pay the tax and, indirectly, the prices they charge and therefore the attractiveness and competitiveness of particular goods and services. For example, if a system charged tax on most food but exempted that which was certified as meeting certain organic food standards, this would be likely to affect significantly the behaviour of consumers and hence retailers and producers.

---

8 E.g. *R (Trailer & Marina (Leven) Ltd)* v. *Secretary of State for the Environment, Food and Rural Affairs* [2004] EWCA Civ 1580, [2005] 1 WLR 1267, in relation to designation as a Site of Special Scientific Interest.

9 Income Tax (Earnings and Pensions) Act 2003; Income Tax (Trading and Other Income) Act 2005.

10 This lies at the heart of the tax breaks for conservation easements in the United States (n. 2).

11 Taxation of Chargeable Gains Act 1992.

12 Value Added Tax Act 1994.

Wealth taxes, based purely on the value of property held, will operate like land taxes to encourage the productive use of all property, but the valuation methods will play a major role. Again, as noted in relation to land, transaction taxes, whether on sale[13] or inheritance,[14] may encourage commercially productive utilisation of property, and the fragmentation of holdings. However, since transaction taxes offer more potential for deferring rather than granting complete exemption from taxation, these create scope for exercising a degree of future control over the land, as in the case of the conditional exemptions discussed later.[15]

Further taxes can be based on specific resources, discouraging or incentivising the use or non-use of such resources. Thus, the additional costs imposed by liability to the aggregates levy (charged on sand, gravel and rock that has been dug from the ground, dredged from the sea in UK waters or imported)[16] benefits biodiversity by discouraging the creation of new mines, quarries and pits for the extraction of minerals. Likewise, the landfill tax (charged on waste being disposed of by landfill)[17] is an incentive to avoid the deposit of waste in the ground, not only reducing the demand for land which can be used as waste sites but also affecting the second income stream that once made it a commercially attractive business to dig large holes to extract gravel, etc. and then fill them with waste. Fuel duties can affect the demand for specific fuels, and the balance between different sources, with consequences for biodiversity a long way back in the supply chain, as with biofuels.[18]

6.2 FORMS OF SPECIAL TREATMENT

As well as there being a wide range of possible forms of taxation, there is a wide range of ways in which special treatment can be given to particular activities or forms of income or expenditure. The most

---

[13] Stamp duty land tax (Finance Act 2003, Pt 4) and stamp duty reserve tax (Finance Act 1999, Sch. 13 and Finance Act 2003, Pt 5); and now the land and buildings transaction tax (Land and Buildings Transaction Tax (Scotland) Act 2013).

[14] Inheritance Act 1984, *né* Capital Transfer Tax Act 1984.

[15] See section 6.3 below.

[16] Finance Act 2001, Pt 2; Aggregates Levy (General) Regulations 2002, SI 2002/761; Aggregates Levy (Registration and Miscellaneous Provisions) Regulations 2001, SI 2001/4027.

[17] Finance Act 1996, Pt III; Landfill Tax (Scotland) Act 2014.

[18] See Campbell and Doswald, *The Impacts of Biofuel Production on Biodiversity* (n. 3).

straightforward is simply a complete exclusion, with property, income or expenditure of a particular sort not taken into account for tax purposes. An example here is forestry in the United Kingdom, where the response to the misuse of the special treatment in relation to income tax has been to remove completely forestry income and expenditure from the tax calculations.[19] Other clear exemptions include agricultural land and woodland being exempt from non-domestic rates,[20] and the exempting of the agricultural value of agricultural property from inheritance tax.[21] Where such complete exemptions are involved and affect large sums (as may be the case with agricultural relief from inheritance tax), those who might benefit go to considerable lengths to ensure that they can, and can demonstrate to the authorities that they do, fall within the relevant conditions to benefit from the exemption. This is one example of the general point that the larger the benefits to be obtained, the more that (potential) taxpayers will do in order to meet the required conditions. Without going as far as providing total exemption, variable tax rates can be used to demonstrate support for or discouragement against activities. An example is the attempt to encourage energy conservation by having a reduced rate of VAT on insulation and other energy saving materials.[22]

A more sophisticated possibility is to offer conditional exemption, as occurs in relation to liability to inheritance tax or capital gains tax for certain kinds of 'heritage property' and is discussed more fully below. Under this scheme, property transactions that would normally incur a tax liability but involve properties which are considered to be of public importance are exempt, but only so long as the taxpayer gives, and continues to fulfil, certain undertakings in relation to the conservation of and public access to the property.

Further scope for special treatment is provided by allowing spending on certain things to be set off against tax liabilities. This can occur either

---

19 Income Tax (Trading and Other Income) Act 2005, s. 11; see section 6.3 below.

20 Local Government Finance Act 1988, Sch. 5; Valuation and Rating (Scotland) Act 1956, s. 7. Local government finance is governed by an appallingly complex patchwork of provisions, some in Scotland dating back over 150 years.

21 Inheritance Tax Act 1984, s. 116.

22 The general application of this reduced rate has been held to be unlawful under EU law which provides for such a concession only in certain circumstances; *Commission* v. *United Kingdom* (C-161/14), CJEU, 4 June 2015; for the background see HM Revenue and Customs, *VAT Manual: VENSAV2000, History of the Reduced Rate* (2015), available at www.hmrc.gov.uk/manuals/vensavmanual/vensav2000.htm.

at the stage of calculating the income or profit from an activity that will be liable to tax, or at the final stage by allowing the total liability arising from some or all sources to be reduced as a result of gifts or expenditure which may be wholly unrelated. As well as the standard rules on expenses, allowances and investment, the scope for the first option to be used to promote desirable behaviour can be illustrated by the slight extension to allow as an expense not only direct expenditure in a trade but also expenditure on medical research for the benefit of workers in the trade in question.[23] Similar extensions could provide an incentive for spending on biodiversity issues that are not directly part of the profit- or income-generating activity[24] but are related to it. The second option of more general set-off has a leading example in the relief given to charitable donations.[25] A similar position in the United States has played a major part in encouraging the use of conservation easements there and determining their precise form, since granting such an easement over a plot of land to a charitable body is treated as a donation to charity of the sum by which the value of the plot is reduced as a result of it becoming encumbered with legal restrictions on its future use.[26]

6.3 TAXATION AND BIODIVERSITY: EXAMPLES

The American experience just related is one example of how tax rules designed to be of general application can influence behaviour in relation to biodiversity.[27] Two examples from UK tax law are worth more consideration to show how aspects of the tax system can have an impact on biodiversity, one unintentional, the other deliberate. The first concerns the taxation of forestry in the 1980s.[28] In recognition of the heavy initial investment and long period before returns can be realised, forestry has

---

[23] Income Tax (Trading and Other Income) Act 2005, s. 87.

[24] And hence stand to be included within the standard calculation of net profit or income.

[25] Income Tax Act 2007, Pt 8.

[26] Daniel Halperin, 'Incentives for conservation easements: the charitable deduction or a better way?' (2011) 74 *Law and Contemporary Problems* 29; see Chapter 5.

[27] For another jurisdiction, see Louisa Mamouney, Jennifer Stace and Caroline Heathcote, 'Incentives for biodiversity conservation in NSW, Australia' (2008–09) 38 *Stetson Law Review* 357, 363–6.

[28] Colin T. Reid, 'Forestry, the law and the environment' in Christopher Rodgers (ed.), *Nature Conservation and Countryside Law* (University of Wales Press, 1996); K. Jan Oosthoek, *Conquering the Highlands: A History of the*

been treated differently from other forms of business in the income tax system. Until the 1980s, the provisions operated so as to allow investment in forestry planting to be set off against income from other sources, thereby greatly reducing tax in that year, whilst at a later stage allowing the income from the sale or harvesting of the trees to escape tax.[29] At a time when the top rate of income tax was 60 per cent, this was an attractive route for high earners seeking to invest their income and reduce their tax liability and a lot of investment went into new planting. The tax relief was available without any consideration of the details of a particular project and on occasions stimulated the development of plantations for which the more direct grant support was not even sought. This support had been thought to be an inevitable element of any commercial plans for forestry and therefore by itself to offer sufficient control over substantial planting, without the need to consider any more direct regulation. The plantations driven by the quest for such short-term financial benefits were not always designed with proper consideration of environmental factors, or even their long-term value. In particular, substantial planting took place in the Flow Country of Sutherland, draining areas of blanket bog which are now recognised as being of exceptional value for biodiversity and as a carbon sink.[30]

Realisation of the adverse, and wholly unintended, impact on biodiversity of the tax incentives produced two responses. The first in 1988 was removal of the tax relief.[31] Commercial woodlands now operate outwith the income tax system, with no tax due on the income but no relief on the expense of establishing new trees. Financial support from public funds is now directed wholly through the grant schemes under which support is provided only to projects which have been individually scrutinised[32] and

---

Afforestation of the Scottish Uplands (Australian National University E Press, 2013) 149–59.

[29] Thomas Lynch, *Taxation of Woodlands in the UK* (W. Green, 1989).

[30] RSPB, *Bringing Life Back to the Bogs* (RSPB, 2011); see the description for potential listing as a UNESCO World Heritage Site: UNESCO, *The Flow Country* (2012), available at http://whc.unesco.org/en/tentativelists/5679/.

[31] Finance Act 1988, s. 65 and Sch. 6; Thomas Lynch, 'Taxation: the end of an era?' 1990 *Scots Law Times (News)* 297.

[32] This may involve an environmental impact assessment; Environmental Impact Assessment (Forestry) (Scotland) Regulations 1999, SSI 1999/43; Environmental Impact Assessment (Forestry) (England and Wales) Regulations 1999, SI 1999/2228.

found to comply with general policy.[33] The pattern of forestry planting in the aftermath of the changes to the tax and grants systems offers an interesting study of the power of economic instruments to change behaviour, but also of the vulnerability of any economic measures to wider economic considerations since an economic downturn meant that there was less of the surplus income to be invested in these or other projects.[34] The second response, coming considerably later, has been substantial effort to reverse the damage done by the inappropriate planting and restore the ecosystem of the bog. Considerable efforts, notably by the Royal Society for the Protection of Birds (RSPB), and support from public funds (including the tax benefits for the RSPB and for donors arising from its status as a charity), have gone in to clearing areas of the trees planted in the 1980s and blocking the drainage that had been introduced.[35] The significance of peatland and active bog systems as carbon sinks is adding extra weight to the biodiversity arguments for such work.

The forestry experience shows the unintentional consequences that aspects of the tax system can produce. A second example demonstrating a deliberate attempt to use the system to provide public goods is provided by the conditional exemption scheme for inheritance tax[36] and capital gains tax,[37] which might provide a model for wider support for biodiversity purposes. To the extent that people are aware of the scheme, under which tax is provisionally waived so long as public access to property of heritage value is provided, it is usually in the context of works of art or stately homes, but its terms do extend to include land of outstanding scenic or scientific interest.[38] Land which is a Site of Special Scientific Interest or a National Nature Reserve is eligible and to gain the exemption an undertaking must be given to manage the land so as to maintain and preserve its character and to secure reasonable public access. The statutory conservation bodies are involved in the acceptance of any land and, in addition to any statutory consents, these bodies must

---

33 These grants are now mostly incorporated within the wider Rural Development Programmes (see Scottish Government, 'Scottish Rural Development Programme', www.scotland.gov.uk/Topics/farmingrural/SRDP.

34 Colin T. Reid, 'Environmental regulation through economic instruments: the example of forestry' (1996) 8 *Environmental Law and Management* 59.

35 See RSPB, *Forsinard Flows* (2012), available at www.rspb.org.uk/reserves/guide/f/forsinard/work.aspx.

36 Inheritance Act 1984, ss. 30–35A.

37 Taxation of Chargeable Gains Act 1992, ss. 257–8.

38 The details of the scheme can be found in HM Revenue and Customs, *Capital Taxation and the National Heritage* (2009), available at www.hmrc.gov.uk/inheritancetax/conditionalexemption.pdf.

be consulted on any development or changes that may affect its character. The public access provided must be reasonable, but that need not involve unlimited access, and access can be further restricted for scientific reasons. In the case of woodland, the maintenance of the land can include felling and replanting, but otherwise the undertaking relates just to the maintenance and preservation of the land, not its enhancement, whilst any work already required under other provisions, such as agri-environment schemes, cannot also be used as the basis of an undertaking. Funds set aside for the maintenance of the site can also benefit from tax relief.

The fact that land is subject to such an undertaking must be published, as may the details of its terms and the access arrangements, which must in any event be available on request.[39] The tax exemption can continue if the new owner agrees to take over the land on the same conditions,[40] but is otherwise lost if the land is transferred or there is any material breach of the undertaking. Where the breach is the result of a genuine error and not a casual, deliberate or repeated disregard of the obligations, the owner may be allowed one opportunity to remedy the situation, if feasible.

This scheme has some features in common with conservation covenants – a long-term obligation owed to a recognised body by the landowner to maintain land in certain condition and capable of continuing through changes of ownership – but is policed by the powerful sanction of losing an often very substantial tax exemption. The undertakings tend to be expressed in fairly general terms, based on maintaining or preserving the character of the site, but more specificity is provided by the need to comply with management plans, agreed with the relevant conservation body and subject to regular updating.[41] Looking at the sites covered by such undertakings,[42] most are there for their scenic and amenity value, for example the grounds of a country house, rather than their scientific interest, but the eligibility conditions are wide enough for this approach to be used for ecosystem purposes.

---

39 See the Tax Exempt Heritage Database at HM Revenue and Customs, 'Tax-exempt heritage assets: database', www.hmrc.gov.uk/heritage/data.htm.

40 It becomes permanent if the land is transferred to any of the bodies listed in Sch 3 to the Inheritance Tax Act 1984 as recipients of gifts for national purposes.

41 See e.g., the undertaking for Wormshill Estate in Kent (unique ID 568; no. 43 on the South East list on the database (n. 39)).

42 Listed on the database (n. 39).

6.4 POTENTIAL AND DRAWBACKS

The tax system undoubtedly has the potential to play a major role in influencing behaviour, but this is most pronounced where there is a very direct connection between the tax signal and the change of behaviour desired. The transition to unleaded petrol in the United Kingdom during the 1980s was greatly stimulated by the introduction of a tax, and therefore price, differential between leaded and unleaded petrol,[43] but this is an example where for most consumers price is the only distinction between functionally equivalent products and where the tax differential made a very clear difference to the final price. This direct connection with the general public as consumers is not likely to be the case in many circumstances where taxes are manipulated for biodiversity reasons, so that landowners and managers are perhaps the more likely focus for adjustments to the tax system. Allowing tax offsets or imposing lower, or no, taxes for land involved in approved conservation schemes, or where owners agree to restriction on the use of land, can influence behaviour, as the experience in the United States has shown.

Despite the potential of the tax system as a means of providing incentives for biodiversity-friendly action, there are also some drawbacks to be considered, both with the approach as a whole and on matters of detail, whilst at the same time there should be awareness to the potential for unintended adverse consequences. One feature of tax mechanisms is that they are of general application, applying in all cases that meet the defined eligibility criteria, without the opportunity to consider the implications in the particular circumstances. To some extent this is an advantage, creating certainty and avoiding the costs of individualised decisions, but it can also be a disadvantage. As the forestry example discussed above shows, if the wider context leads to people acting in ways that were unforeseen and are undesirable, it is not usually possible

---

[43] The differential, introduced in 1987, reduced the price of unleaded petrol. The potential saving in petrol costs helped to overcome the inertia arising from the minor hassle and cost of making the adjustments needed to allow older cars to run on unleaded petrol and influenced choice when buying a new car. This created a virtuous circle of increasing demand for and greater availability of unleaded fuel, overcoming another constraint on the wider adoption of the new fuel. In one author's recollection, local supply issues and fuel price as a spur to the conversion of existing vehicles (expressly excluded from the study) were much more significant than is suggested in one economic analysis of this process: Paul Stoneman and Giuliana Battisti, 'The role of regulation, fiscal incentives and changes in tastes in the diffusion of unleaded petrol' (2000) 52(2) *Oxford Economic Papers* 326.

to disapply the rules for certain cases but not others – the rules have to be rewritten. A possible way round this is to link tax measures with other schemes which do take a site-specific approach, for example, setting eligibility criteria based on designation under a statutory conservation scheme or inclusion in another conservation scheme, such as receipt of the more advanced agri-environment payments.[44] Similarly, treatment based on the nature of the taxpayer, such as exempting land held by bodies that are charitable or are designated on some other basis, can be used to define the limits of the provisions.

At a more detailed level, the precise phrasing of any tax rules will be important. Thus in theory, the fact that it is land used for agriculture and woodlands which is exempt from non-domestic rates[45] but not land dedicated wholly to nature conservation, might create a difficulty. In other jurisdictions a range of other means are used to identify the land that is to benefit from special treatment. Agricultural exemptions are common, but may[46] or may not[47] include conservation management within this category. In relation to land taxes, Illinois applies a reduced rate for open-space, that is areas of at least 10 acres used for at least three years exclusively for maintaining or enhancing natural or scenic resources, or for promoting conservation of natural resources; similar provision applies to vegetative filter strips by watercourses and land

---

44 E.g. in relation to inheritance tax, see HM Revenue and Customs, *Agricultural Purposes: Qualifying Uses of Land: The Habitat Scheme*, IHTM24065 (2015), available at www.hmrc.gov.uk/manuals/ihtmanual/ihtm 24065.htm. Such arrangements may, however, raise concerns over double-counting and a lack of additionality, as discussed below.

45 See n. 5 above.

46 E.g. in Texas where wildlife management falls within the open-space/ agricultural exemption; Susan Combs, *Property Tax Basics: Texas Property Tax*, Publication #96-1425 (Texas Comptroller of Public Accounts, 2014) 16. In Louisiana, marsh land is treated as in the same class as agricultural, horticultural and timber lands; Louisiana Tax Commission, *Rules and Regulations*, ch. 27, available at www.latax.state.la.us/Menu_RulesRegulations/RulesRegulations. aspx.

47 E.g. in Queensland where land used solely for conservation purposes is expressly excluded from the agricultural exemption; Queensland Government, *Overview of Land Tax* (2015), available at www.osr.qld.gov.au/land-tax/about-land-tax/index.shtml and Queensland Government, *Exemption for the Business of Agriculture, Pasturage or Dairy Farming – Forestry*, LTA053.2.1 (2010), para. 12, available at www.osr.qld.gov.au/legislation-rulings/public-rulings/land-tax/ lta053-2.shtml.

meeting the requirements for conservation stewardship.[48] In similar contexts many other jurisdictions include an exemption not just for agricultural land but also for land designated under various biodiversity regimes[49] or subject to specified forms of conservation agreements.[50] These ways of categorisation can be supplemented by rules based on the nature of the taxpayer as a charitable or not-for-profit body.[51]

In practice, though, where there are not large sums at stake and no commercial activities appear to be escaping tax when they might be expected to incur liability, the likelihood is that technical distinctions between agricultural or forestry uses and conservation land will be avoided by the fact that there may well be some grazing or woodland cover on the relevant land, even though it is not the main use.[52] Moreover, it is understood that the practice of assessors is to assume that rural land is agricultural unless it is clearly being used for another commercial purpose.[53] Greater certainty can be obtained by linking tax exemptions to other conservation measures, which also enables the fine details of eligibility to be determined in those contexts rather than by tax officials.

A more fundamental drawback with using tax relief of any sort is that this can only be effective where there is a tax liability to be relieved. This affects both those who may benefit and those who may not, and raises questions over the appropriate use of the public funds involved.[54] Where the tax measure takes the form of setting-off expenditure[55] against profit

---

48 Illinois Department of Revenue, *The Illinois Property Tax System: A General Guide to the Local Property Tax Cycle* (Illinois Department of Revenue, 2014) 27, available at www.revenue.state.il.us/publications/localgovernment/ptax1004.pdf.

49 E.g. nature reserves or national parks in South Africa, except for those areas within these developed or used for commercial, business or residential purposes; Municipal Property Rates Act 2004, s. 17.

50 E.g. Government of Western Australia, Department of Finance, Office of State Revenue, *2014–15 Land Tax Brochure*, 4.

51 *Ibid.*

52 The ending of the business rates exemption for 'shootings and deer forests' under the Land Reform (Scotland) Act 2016 (ss. 74–6) will draw attention to definitional issues.

53 Personal communication with local assessors in Scotland.

54 The sums foregone through tax relief are as much a drain on the public purse as direct expenditure in subsidies.

55 Or certain gifts, as in the case of donations of conservation easements in the United States.

or income from other sources, then only those who are making substantial profits or income will benefit; where it involves a total exemption, that is beneficial only to those who would otherwise have to pay the tax. It was wealthy investors from show business and other fields, not those struggling to run rural estates and farms as viable businesses, who could take most advantage of the special treatment of forestry in the United Kingdom before 1988; it is only those inheriting property worth a substantial amount and thus attracting inheritance tax who benefit from the conditional exemption for heritage property. The public funds involved in giving the tax relief are thus going to the benefit of the wealthy, rather than directly to support the management activities which may be benefiting biodiversity. Indeed, the low intensity of use and limited commercial returns which are features of much land of high biodiversity value reduce the likelihood of gaining benefits from any tax relief predicated on the high value of the land attracting liability for either a property or transaction tax. Moreover, relief from taxes based on inheritance or other transactions will be of no use unless such transfers take place, which with a family-owned estate may be only once in a generation and even less frequently for property held by trusts or corporations.

A separate issue relating to the appropriate use of public funds concerns paying people for doing what is already required of them by other means. This is a manifestation of the broader concern over additionality which affects many approaches,[56] but has a particular impact since public funds are involved. Thus, the conditional exemption from UK inheritance tax expressly provides that management activities already required under other schemes cannot be the basis for an undertaking that confers tax relief.[57] A danger here is that as the general rules for legally designated conservation sites have evolved from purely negative requirements to requiring more positive management of the land for conservation purposes, the risk of an overlap increases. Moreover, in practice land managers have been faced with a range of different schemes based on financial incentives, which run for different periods and offer more or less benefit than tax relief depending on the individual circumstances of the taxpayer. There may thus have to be complex calculations to ensure there is no double-counting and to work out the most advantageous approach for the taxpayer.

---

[56] See Chapter 2, section 2.5.

[57] HM Revenue and Customs, *Capital Taxation and the National Heritage* (n. 38).

Three further pervasive issues arise in the taxation context. The first is valuation. Where an exemption is directly linked to a specific tax liability arising from the property affected, such as the conditional exemption from inheritance tax, the position is fairly straightforward, but where there is some set-off across a broader canvas things become more difficult. This is a point highlighted by the US experience of conservation easements, with concern that the reduction in the value of land attributed to the restrictions imposed by an easement, and thus the sum being treated as a charitable gift to be set off against other tax liabilities,[58] may be overstated, to the benefit of the taxpayer and detriment of the public purse.[59] The second is measuring the effectiveness of any tax measure and calculating whether it is providing value for money, where again the US experience is relevant.[60] Would the sums of money spent on tax relief achieve more if devoted to direct payments for land management to support biodiversity? Finally, the effect of any tax measures will be felt on a haphazard basis across the country, depending on where and whether a taxpayer can benefit from the relief, rather than contributing to either a landscape-scale approach or a coherent conservation strategy devised to focus effort where most appropriate.

One advantage of employing the tax system is that it offers ready-made procedures for enforcement and sanctions, but monitoring effectiveness may be more problematic. This is especially so since the biodiversity gains which are being sought in exchange for the tax relief may require continuing management over, and will only be confirmed after, several years. Giving full tax relief today in exchange for benefits that may or may not be delivered at a future date is not an evenly balanced trade, and monitoring has been identified as a significant issue in relation to the conservation easements in the United States entered into in exchange for charitable tax relief.[61] Where a continuing commitment is involved, as with the conditional exemption from UK inheritance tax, even though the

---

58 There is also the issue of the assessment of the current value of the land, which may affect land taxes, e.g. Alaska expressly provides that full account is to be taken of any conservation easement in valuing the land; Alaska Statutes 29.45.062.

59 Anna Vinson, 'Re-allocating the conservation landscape: conservation easements and regulation working in concert' (2006–07) 18 *Fordham Environmental Law Review* 273, 280–1; Zachary Bray, 'Reconciling development and natural beauty: the promise and dilemma of conservation easements' (2010) 34 *Harvard Environmental Law Review* 119, 145–8.

60 *Ibid.*

61 See Chapter 5, section 5.6.

tax authorities may have powers to enter and inspect the land[62] they may not have the expertise to assess compliance with ecological criteria. In such circumstances, linking the tax benefits with continuing compliance with a plan or programme supervised by a more appropriate body (for example, a management agreement with a statutory conservation body) may offer a way forward.

6.5 HANDLING OF REVENUES

It is not just the structure and details of the taxes that are being raised that can affect conservation, but also the ways in which the money collected is to be used. Obviously, the extent to which any public revenues are devoted to biodiversity purposes is important but there are two specific issues which merit some attention in thinking about the role of taxation for biodiversity, one that applies generally to environmental taxation and one more distinct. The first is whether any revenue raised should be dedicated to specific purposes, so that taxes raised from activities causing environmental harm are devoted to environmental purposes. The second is compensating the bodies affected where a source of revenue is restricted for biodiversity reasons, for example, the amount of land tax raised by a local authority because of exemptions for extensive nature reserves in its area.

6.5.1 Hypothecation

The dedication of the revenue from a specific tax to a specific purpose, known as 'earmarking' or 'hypothecation',[63] is an issue that has often been raised in the context of environmental taxation. Whereas the general pattern is for the proceeds of all taxes to be gathered into the one consolidated fund from which all the activities of government are financed, under a hypothecated system the sums raised from a designated tax are available for use only for the designated purpose. An explicit connection between the taxes paid and the benefits they provide is seen

---

62 The information and inspection powers in Sch. 36 to the Finance Act 2008 are very broad.

63 Stephen Bailey, *Public Sector Economics: Theory, Policy and Practice* (2nd edn, Palgrave, 2002) 231–42; Andy Lymer and Lynne Oats, *Taxation: Policy and Practice* (19th edn, Fiscal Publications, 2012) 55–6; Antony Seely, *Hypothecated Taxation*, House of Commons Library Standard Note, SN01480 (2011), available at www.parliament.uk/briefing-papers/SN01480.

as a means of gaining public accountability and approval for new, increased or continued taxes, but has not been widely adopted in the United Kingdom.[64] The widespread adoption of such an approach not only complicates fiscal management but also leaves the different strands of the public finances vulnerable to fluctuations in public support and the receipts from particular sources, whilst the supposed benefit of greater transparency and accountability is undermined by the limited short-term flexibility across so many areas of public spending. A further danger is that if a tax is also used to discourage undesirable activity, the greater the success in discouraging the activity, the less the revenue generated and thus available for the hypothecated purpose; for example, if the health service were largely supported by duties on tobacco and alcohol, then successful public health campaigns reducing consumption would have the effect of reducing the resources available for the service, long before the financial savings resulting from the health benefits came into play.[65]

There is a connection here with what in UK parlance are known as tax credits, relieving the taxpayer of a liability that would otherwise accrue. Such credits in effect involve money that would otherwise fall into the general revenue receipts being allocated for a specified purpose, but in this case the money is being provided directly by the taxpayer for the recognised purposes rather than being raised as taxes and then distributed. When operated at a broad level, there may be no close connection in terms of purposes such as is the essence of hypothecation, for example, allowing charitable donations across the whole range of charitable purposes to be set off against income tax which contributes to the

---

[64] There seems to be a growing willingness to adopt this approach, e.g. the announcement in the July 2015 Budget that: 'From 2020–21 the government will spend all of the revenue raised from [vehicle excise duty] in England on the English Strategic Road Network' (HM Treasury, *Summer Budget 2015*, HC 264 (2015–16) para. 2.146) and the introduction of the apprenticeship levy the proceeds of which will be spent on supporting apprenticeships (HM Treasury, *Spending Review and Autumn Statement 2015*, Cm 9162 (2015) 45. The TV licence fee which supports the BBC can also be seen as a hypothecated tax. See Bailey, *Public Sector Economics* (n. 63) 231–2 and Seely, *Hypothecated Taxation* (n. 63).

[65] OECD, *OECD Economic Survey: United Kingdom, 1999–2000* (OECD, 2000) 75. There can be argument over whether the increase in early deaths arising from smoking actually results in a financial benefit to the health and related services by avoiding the costs of pensions and the long-term care of the elderly; W.K. Viscusi, *Cigarette Taxation and the Social Consequences of Smoking*, Working Paper No. 4891 (National Bureau of Economic Research, 1994).

general public finances. Such tax credits can, however, operate at a much more narrowly focused level. In the United Kingdom, the outstanding example of this is the Landfill Communities Fund, whereby the operators of landfill sites can claim credits against their liability to the landfill tax (levied per tonne of waste disposed of by way of landfill) for the contributions they make to registered environmental bodies and which are used for one of the purposes authorised under the scheme.[66] The operator can claim credits amounting to 90 per cent of the payments it has made, but subject to a maximum set as a proportion of its total liability for the tax, initially set at 20 per cent but since 2003 reduced to about 6 per cent.[67] In this way the receipts from the landfill tax are not dedicated to a specific purpose, but some of the money that would otherwise have been collected is.

Detailed safeguards are in place to ensure that the payments are used for environmental purposes and not to benefit the operator. The beneficiary of the payment must be a not-for-profit body, must not be controlled by a local authority or landfill operators and must be enrolled with Entrust, the body established to administer the scheme. The funds must be used for one of a limited number of purposes, namely the reclamation of, or reduction in pollutants on, land, or in the vicinity of the landfill site work on a public park or other public amenity, on buildings or structures of architectural or religious interest or on the conservation or enhancement of biological diversity. The limitation to activity in the vicinity of a landfill site means that the credits given are being used to provide some benefit to the communities most affected by the presence of a landfill site, a feature which serves as an attraction to operators who cannot benefit directly from any work paid for but whose image in the community can be improved by support for local projects. Similar schemes could be developed to direct some of the resources which would otherwise be raised in tax from other environmentally harmful activities towards supporting appropriate environmental enhancements.

---

66 Finance Act 1996, Pt III, especially s. 56; Landfill Tax Regulations 1996, SI 1996/1527, Pt VII; see HMRC, *A General Guide to Landfill Tax*, Notice LFT1 (May 2012); Bailey, *Public Sector Economics* (n. 63) 238–9. Since 2015, the landfill tax in Scotland has become a responsibility of the Scottish Government but a broadly similar pattern remains; Landfill Tax (Scotland) Act 2014.

67 Set at 5.6 per cent for 2012–13, amounting to £78.1 million; Landfill Tax (Amendment) Regulations 2012, SI 2012/885; HM Revenue and Customs and HM Treasury, *Overview of Tax Legislation and Rates* (March 2012) para. 1.46.

6.5.2 Distributive Effects Among Taxing Bodies

Where taxes are affected by biodiversity and raised on a regional or local basis, biodiversity policy will have a distributional effect between the revenue raising bodies that may have to be addressed. An authority, a significant proportion of whose area is dedicated to conservation, is likely to be able to gather less revenue from taxes than one where more of the land is being more intensively exploited. This may be as a result of tax policy specifically exempting from liability any property or activity devoted to conservation, or may arise simply because biodiversity policy prevents the development of land in the more intensive ways that generate tax income. Whatever the background, the consequence is that by supporting conservation in its area, an authority risks reducing (or limiting the expansion of) its own revenues, producing an incentive not to support biodiversity measures. Equally, where the strong biodiversity interest is the result of geographical features or historical patterns of development concentrating economic growth in other areas, the authority in an area rich in biodiversity may be financially worse off and therefore have particularly strong financial incentives to encourage economically profitable land uses over conservation.[68] These considerations apply all the more if the authority concerned is also responsible for paying some of the costs of conservation activity within its area.

In order to prevent such considerations distorting policy, it may therefore be desirable to introduce a system of fiscal transfers between authorities, in effect to compensate those who are bearing a disproportionate share of the national biodiversity burden. Such mechanisms may be particularly significant where there is scope for conservation initiatives at national, regional and local level, since otherwise there may be local resentment at the negative impact at the local level resulting from conservation decisions taken at higher levels of government. A system of such transfers operates in Brazil in the form of the *ICMS Ecológico*, where a proportion of the revenue raised by states from the tax broadly equivalent to VAT[69] is shared in a way that takes account of ecological

---

[68] This is the expression on a national scale of the arguments at international level on the unfairness of the developed countries which have severely damaged the global environment in achieving economic development imposing environmental constraints on developing countries that limit their ability to develop; cf. Principle 7 of the Rio Declaration on Environment and Development, UN Doc. A/CONF.151/26 (June 1992) vol. I, (1992) 31 ILM 874.

[69] Imposto sobre Circulação de Mercadorias e Serviços (ICMS).

circumstances.[70] The law requires the state tax to be allocated between municipalities, predominantly on the basis of where the revenue is raised, but an 'ecological' element is also taken into account for the benefit of those municipalities heavily affected by conservation designations and who are thus unable to develop the economic activity that will generate revenues.[71] The proportion of the revenue allocated in this way and the details of the allocation basis vary between states, but the principle is clear and means that the tax system provides some incentive (or at least reduces the disincentive) to maintain and expand the protected areas, rather than the opposite. The same idea of fiscal transfers also operates in Portugal,[72] and researchers have developed a detailed proposal in Germany.[73] Such studies show the many variables that could be used in determining what is to be used as the basis for any transfers (simple area of designated land or a more sophisticated measure?), the interaction with other transfers of funds, and the complexity of working out the relative winners and losers under such systems.

In the UK context, there is no equivalent mechanism but recent reforms in England which are moving in the opposite direction provide a useful example of the interaction between conservation and other policies. The systems of local government are different in each part of the United Kingdom[74] and the financial arrangements are exceedingly complex, with the bulk of their funding coming through central government.[75] Property taxes (rates) are levied on domestic and non-domestic

---

[70] Irene Ring, 'Integrating local ecological services into intergovernmental fiscal transfers: the case of the ecological ICMS in Brazil' (2008) 25 *Land Use Policy* 485.

[71] In one case 90 per cent of a municipality's area is designated to protect water supplies and the remaining 10 per cent for biodiversity; *ibid.* 488.

[72] Rui Santos and others, 'Fiscal transfers for biodiversity conservation: the Portuguese local finances law' (2012) 29 *Land Use Policy* 261.

[73] Irene Ring, 'Compensating municipalities for protected areas: fiscal transfers for biodiversity conservation in Saxony, Germany' (2008) 17 *GAIA – Ecological Perspectives for Science and Society*, Supplement 1, 143.

[74] There are variations within England and London has a quite separate structure.

[75] In Scotland 85 per cent, Wales 80 per cent and England 53 per cent; Scottish Government, *Local Government Finance Settlement* (2011), paras 6 and 21, available at www.scotland.gov.uk/Topics/Government/Finance/spfm/locgov fin#a6; Welsh Government, *Local Government Settlement* (2015), available at http://gov.wales/topics/localgovernment/finandfunding/settlement/?lang=en; Department for Communities and Local Government, *Local Government Resource Review: Proposals for Business Rates Retention: Consultation* (2011)

property; for non-domestic property the rate is set by and the revenue goes to central government; for domestic property ('council tax') both elements are in the hands of the local authority.[76] Reform in England is seeking to incentivise authorities to promote economic growth by allowing authorities to retain a proportion of the non-domestic rates that are raised within their area, so that there is a direct benefit to them in achieving such growth.[77] The legislation that has now been passed[78] makes provision for this (with various limitations and qualifications). At no stage has the impact of this measure on conservation been a topic for discussion; indeed, the whole point is to give higher priority to promoting local economic growth rather than other objectives. Thus, although in taking planning and other decisions, authorities are bound by their biodiversity duty[79] and conservation will always be a material consideration, there is a now an additional countervailing factor to be added to the balance in favour of development. The legislation does allow for special treatment for some areas of land, but this is designed to take account of the added incentives and activity in enterprise zones and is not structured so as to enable the presence of a significant area of protected land to be accommodated.[80]

Given the existing pressures in favour of development, the fact that only a small proportion of non-domestic rates will be retained and the many factors taken into account in determining the allocations from central to local government,[81] this reform is not likely by itself to have a

---

10. The structure and functions of local government in Northern Ireland are too different for a comparative figure to be meaningful.

[76] Although there are statutory controls and political agreements with central government that have, for example, meant that the rate in Scotland has been frozen for several years; Scottish Government, *Local Government Finance Settlement* (n. 75) para. 6.

[77] Department for Communities and Local Government, *Local Government Resource Review: Proposals for Business Rates Retention: Consultation* (2011) ch. 2; HM Treasury, *Spending Review and Autumn Statement 2015* (n. 64) 58.

[78] Local Government Finance Act 2012, s. 1 and Sch. 1.

[79] The duty to 'have regard, so far as is consistent with the proper exercise of [its] functions, to the purpose of conserving biodiversity'; Natural Environment and Rural Communities Act 2006, s. 40.

[80] Local Government Finance Act 2012, Sch. 1 para. 8 and Explanatory Note, para. 72.

[81] See Department for Communities and Local Government, *Business Rates Retention: Policy Statement* (2012), available at www.gov.uk/government/uploads/system/uploads/attachment_data/file/14939/Business_rates_retention_policy_statement.pdf.

great impact on authorities' balancing of conservation against development. Nevertheless, it serves as a useful example of how the details of complex systems apparently far removed from direct conservation can provide incentives that steer activities in favour of or against biodiversity. Since the reform here points in the same direction as the already well-established priorities for economic growth, it does not attract much attention. If, on the other hand, the reform had proposed increased allocations to local authorities on the basis of the proportion of their area dedicated to conservation, then the influence that such measures can have on setting priorities might have been a lot more obvious (and the political discussion more heated).

7. Transferable development permits, quotas and impact fees

This chapter considers briefly a number of other mechanisms which can contribute to biodiversity conservation and include a 'privatised' element. As an alternative to the direct regulation applied through land use planning laws, schemes involving transferable permits can create a system where trading plays a part in the allocation of development within an area. Indirect impacts on biodiversity can also be influenced by the trading of other rights, such as to water, whilst fishing and hunting quotas apply a more direct market-based approach to the future of fauna. Collecting impact fees when land use is changed is a further means by which contributions from the private sector can be gathered to make up for the adverse consequences of development activities.

Such mechanisms share certain features with those discussed earlier in the book, justifying their consideration here. Again there is a move away from direct 'command and control' regulation to a system where individual actors have more say over what exactly is going to happen. The benefits or risks to biodiversity are the cumulative outcome of many individual decisions made within a broad framework and depend upon the scale and details of the trading or other activity, rather than being the result of a detailed regulatory programme. This means that many of the pervasive issues noted in Chapter 2 arise, especially questions over valuation, transaction costs and the lack of control over the precise outcomes that will be achieved. The regulatory function of the state is not removed altogether, but moved back a step so as to concentrate on providing frameworks within which there is more scope for individual initiative and preference.

7.1 TRANSFERABLE DEVELOPMENT PERMITS

Loss of habitat to urban development or agriculture (or the intensification of existing 'eco-friendly' agriculture) is a major contributor to biodiversity loss throughout the world. Accordingly, techniques to limit the loss of habitat by regulating changes in land use can play a significant part in conservation. Conventionally, the legal approach has been to use various systems of land use planning, but again this is an area where a more 'privatised' approach can be taken. Market-based mechanisms such as tradable development permits to determine the location and scale of land use changes can be used. These differ from biodiversity offsets since they may not try to calculate the direct impact on biodiversity but operate more indirectly, concentrating instead on classes of activity which have major consequences for biodiversity, even though this is not their explicit focus. The same can be said of various mechanisms which are used to manage water or other natural resources, where again an impact on biodiversity is a significant but incidental effect but not the central focus of the regime.

Debates over the extent to which landowners are, and should be, free to do what they wish with their land are an enduring feature of debate on both property law and social policy. Views on the nature of ownership change over time,[1] and remain contested.[2] One major strand of recent discussion is how far landowners should be viewed as subject to obligations based on stewardship,[3] whilst renewed interest in the public trust doctrine is exploring another way in which the rights of owners are not absolute but qualified.[4] Regardless of such discussions, the practical task of regulating land use has generally fallen on some mechanism for land use planning. These may well include more or less powerful forms

---

1 Sean Coyle and Karen Morrow, *The Philosophical Foundations of Environmental Law: Property, Rights and Nature* (Hart Publishing, 2004).

2 David Grinlinton and Prue Taylor (eds), *Property Rights and Sustainability: The Evolution of Property Rights to Meet Ecological Challenges* (Martinus Nijhoff, 2011).

3 William N.R. Lucy and Catherine Mitchell, 'Replacing private property: the case for stewardship' (1996) 55 *Cambridge Law Journal* 566; for a detailed analysis of different concepts of stewardship, see Emily Barritt, 'Conceptualising stewardship in environmental law' (2014) 26 *Journal of Environmental Law* 1.

4 Earlier articles, such as Joseph Sax, 'The public trust doctrine in natural resource law: effective judicial intervention' (1970) 68 *Michigan Law Review* 471, appear now to be stimulating greater academic interest and even litigation; see e.g. Mary Turnipseed and others, 'Reinvigorating the public trust doctrine: expert opinion on the potential of a public trust mandate in U.S. and international environmental law' (2010) 52 *Environment: Science and Policy for Sustainable Development* 6; Raphael Sagarin and Mary Turnipseed, 'The public trust doctrine: where ecology meets natural resources management' (2012) 37 *Annual Review of Environment and Resources* 473; Mary Wood, *Nature's Trust: Environmental Law for a New Ecological Age* (CUP, 2013); Edward Fitzgerald, 'The Alaskan wolf war: the public trust doctrine missing in action' (2009) 15 *Animal Law* 193.

of forward planning, but at the heart of such mechanisms there is usually some form of direct regulation requiring prior approval before certain forms of land use change can lawfully take place.

There are many variations, covering a range of 'traditions of spatial planning',[5] with stronger or weaker connections to economic planning and more or less reliance on strict zoning as opposed to discretionary permitting.[6] In the absence of a nationalisation of land, such systems tend to straddle the 'privatised' and regulatory boundary. Whilst landowners retain most control over their property, they are deprived of the full right to change the use of their land as they wish, since the state's permission is necessary before the value of the land can be released by converting it to the most profitable uses. On the other hand, the law does not impose a pattern of land use that must be followed nor insist that every proposal which is approved is actually carried through. The system is predominantly permissive and reactive, responding to individual proposals for development and leaving their progress in the hands of the parties directly involved. The issue that has arisen is whether there is scope for a more market-based approach to be applied to the allocation of what is judged to be the acceptable level of development within the general framework prescribed by the regulatory authority.

The emergence of various schemes for transferable development permits has been one answer to that, although their role and operation must be viewed in their wider legal context. Whereas in the United Kingdom, the town and country planning system created in the 1940s and still in force simply removed the rights of landowners to develop their land as they please (sometimes referred to as a 'nationalisation of development rights'), in the United States such a diminution of the property rights of landowners has been deemed unacceptable without compensation.[7] This led to a search for alternative means to enable owners to realise the development value of their land without doing so to the detriment of the

---

[5] European Commission, *EU Compendium of Spatial Planning* (European Commission, 1997) 33–7.

[6] 'Although sometimes deemed a single conception, planning and zoning do not cover identical fields of municipal endeavor. While municipal planning embraces zoning, the converse does not hold true. They are not convertible terms. Zoning is not devoid of planning, but it does not include the whole of planning. Zoning is a separation of the municipality into districts, and the regulation of buildings and structures, according to their construction, and the nature and extent of their use, and the nature and extent of the uses of land.' Heher J in *Angermeier* v. *Borough of Sea Girt* (1958) 27 NJ 298, 308, Supreme Court of New Jersey.

[7] See Chapter 1, section 1.3.

ecological or cultural benefits of leaving their land in its current state or only slightly modified. Transferable development rights achieve this by allowing a developer elsewhere to do more than would normally be permitted (for example, to build at a greater density) by acquiring the development rights from the land which is subject to tighter restrictions, rights which cannot be directly exercised but still retain value because of the potential for them to be transferred in this way.

Such schemes require considerable preparation. In the first place there must be a framework for controlling development, a series of legal restrictions on how land in certain areas can be used. It is then necessary within that wider area to identify, on the one hand, a 'supply area' where development is tightly constrained, creating the position where the owners' theoretical rights to develop cannot be used *in situ*, and on the other a 'demand area' where development over and above what would normally be allowed will be permitted so long as the landowner has acquired additional development rights. The owners in the supply area are thus not being deprived of their right to develop but those rights can only be exercised in other locations, requiring a transfer (at an appropriate price) to those wishing to maximise development in the areas where this is allowed. The idea is that within the neighbourhood as a whole, the same level of development as would normally be allowed can take place, but it is concentrated in the areas where this is appropriate, protecting the areas with features thought worthy of conservation. As with biodiversity offsets, areas of particular conservation value can be given stronger protection and excluded from any scheme for transferable rights.

The design challenges in creating such a scheme are considerable and its operation requires a level of regulatory input and supervision that contradicts the claim for market-based instruments to be less intrusive and demanding of regulatory effort. Questions over valuation and transaction costs arise, whilst a means has to be found to secure the necessary long-term restrictions on land in the supply area once the development rights relating to plots have been sold. Moreover, experience suggests that few such schemes work in a way that generates many transfers or effective results.[8] Indeed, this has led to the innovative suggestion that such schemes might work better if reversed, with additional rights having to be bought by landowners wanting to develop at below a set density

---

[8] Ari Bruening, 'The Siren song: the problems with transferable development rights programs and how to fix them' (2008) 23 *Journal of Land Use* 423; John Corkindale, 'Using transferable development rights for land use planning purposes' (2013) 25 *Environmental Law and Management* 223.

rather than above, providing an incentive to avoid low density developments in favour of more intensive ones, which use less open land and can be more sustainable in resource and energy use.[9]

As noted above, though, it is important to consider any schemes in their context. The specific concerns of US law at federal and State levels, and the difficulties (or expense of compensation) of establishing strong regimes for direct control of land use have played a major part in shaping the American examples most commonly discussed. That context does, however, highlight the fact that in all cases, a system based on transferable rights rests on the existence of a regulatory background which can prevent the full exercise of rights where approval has not been obtained. Such regulatory control could serve not as part of the standard structure of land use controls, but as a supplementary scheme more directly focused on habitat. This would only be possible where there was the legislative willingness to impose, in addition to the standard development control system, a further layer of control, albeit one that operates on market-based principles, and to devote the resources needed for the design, administration and enforcement of the scheme. This might be imagined where there are strong competing claims on a limited form of habitat.

An example might be based on the recognition that a particular sort of habitat is in particular demand, both for its biodiversity value and its development potential, for example, coastal sites. The decision could be taken that the economic and social gains are such that it is acceptable for 20 per cent of the available habitat to be lost (or similarly that the nest sites of a certain percentage of a bird population could be damaged), but that the location of such development should not be dictated by the state. A specific scheme could be introduced, operating over and above the standard land use planning system, so that in addition to receiving planning permission in the usual way (thereby maintaining control over any project that is inappropriate in the normal run of things), any development would also need sufficient 'coastal development credits' to cover the scale of the development, one credit per metre of coast affected. These credits could initially be given to all coastal landowners on the basis of one for every 5 metres of coast held, but transferable to other owners. In this way the maximum extent of development matches the 20 per cent loss envisaged as acceptable, but the location of this would depend on the outcome of trading between coastal landowners. Thus, although control over development is maintained, one key element is left

---

9 Bruening, 'The Siren song' (n. 8).

to be determined by market allocation rather than resting wholly on the decisions of the relevant official regulatory authority.

The systems described above deal with immediate development rights in a specific locality. Schemes operating at a different level might also be imagined. A project in Germany undertook a hypothetical experiment based not on landowners trading the right to develop on specific sites, but on municipalities trading between themselves as a means of allocating the right to develop on greenfield sites.[10] The underlying problem identified as requiring attention was that of urban sprawl, with development taking place on the agricultural and other land around the edges of urban areas rather than making use of brownfield land or seeking to fill gaps within existing built-up areas. The use of greenfield sites obviously has negative consequences for biodiversity, but will often provide an easier and cheaper option than development on sites that may be contaminated from previous development or have various constraints because of adjacent buildings and uses. The experiment set a cap on greenfield development significantly below current rates, allocated development rights between municipalities and then allowed their representatives (in one run, existing planners and in another, students) to trade between themselves when deciding where to locate proposed development. The cost of the additional certificates needed to concentrate building on greenfield sites had to be considered against the additional cost of building within the urban areas, whilst the income from selling surplus certificates could contribute to the additional costs arising from the extensive use of urban plots which was creating that surplus.

This experiment is interesting for several reasons. First, it shows the potential for transferable rights to operate on a larger scale than individual development plots and between authorities, not individuals or commercial bodies. Secondly, the results suggested that such a system may not necessarily achieve the expected efficiency gains. Thirdly, it revealed that the students responded more effectively to taking a purely economic approach to the decisions before them, whereas the experienced planners were more influenced by other considerations. Two lessons can be taken from this. The first is that it highlights the extent of the change in thinking that may be required in moving from existing approaches to market-based instruments; the students found it easier to concentrate solely on the economic balance created by the trading

[10] Jens Müller and others, *A Field Experiment with Tradable Development Rights in Germany* (2010), available at www.webmeets.com/files/papers/WCERE/2010/819/Spielraum_25_06_2010.pdf.

scheme. The second is that mechanisms based on economic models may not in practice (or even in tests) always be operated on the basis of purely economic considerations alone. The experienced planners continued to be influenced by the many other factors which to them contributed to sound planning decisions. Depending on one's view of trading models, this can be seen as either a weakness (undermining the fundamental logic of seeking optimal allocation on an economic basis that reflects all costs) or a strength (revealing that trading structures need not operate in a blinkered way but can be moderated by wider considerations).

7.2 WATER RIGHTS

One area where transferable rights are more widely used, with significant impacts for biodiversity, is in relation to water resources. An interesting feature here is the emerging acceptance that the needs of the natural environment should be recognised in the allocation of the available water, but this can lead to significantly different responses. Examples from South Africa and Australia illustrate two contrasting ways of giving effect to the recognition of nature's needs.[11]

In South Africa, environmental needs are taken into account before the more detailed allocation process begins, sitting outside the allocation system that recognises private rights. A fundamental feature of the water resource management framework is protection of the 'Reserve', which is water required both 'to satisfy basic human needs' and 'to protect aquatic ecosystems in order to secure ecologically sustainable development and use of the relevant water resource'.[12] The linking of ecosystems with the use of the water resource shows that the approach being taken is not wholly ecocentric, but environmental needs are being taken outwith the

---

[11] Lee Godden, 'Water law reform in Australia and South Africa: sustainability, efficiency and social justice' (2005) 17 *Journal of Environmental Law* 181; Louis Kotzé and Rebecca Bates, 'Similar but different: comparative perspectives on access to water in Australia and South Africa' (2011–12) 15 *University of Denver Water Law Review* 221.

[12] National Water Act (No. 36 of 1998), ss. 1, 16–18. There is tension between the constitutional rights to a healthy environment and to water (Constitution of the Republic of South Africa, 1996, arts 24 and 27) and over the status of the statutory, but not constitutional, provision for the Reserve; Louis Kotzé, 'Phiri, the plight of the poor and the perils of climate change: time to rethink environmental and socio-economic rights in South Africa?' (2010) 1 *Journal of Human Rights and the Environment* 135.

competition between users and given priority over individual rights, subject to satisfying the human right to water.[13]

In Australia too, the environment has a prominent place in the various water allocation processes, being recognised as having claims on the resource not only equal to but having priority over those of other water users.[14] The approach, though, can be to make nature an integral part of the resource management process rather than as something sitting outside it. One way of achieving this is the appointment of an Environmental Water Holder, as in the State of Victoria,[15] whose role is to hold and manage water entitlements to maintain and enhance the environmental water reserve.[16] The water management structures involve strong elements of market-based methods, but the legislative and planning framework recognises nature's right to be a (privileged) participant in the allocation negotiations. The environment becomes another 'player in the game', represented through an 'agent', engaging in the allocation process which is predominantly about distributing private rights. The interests of the natural environment are thus incorporated into a privatised framework, rather than being either overlooked or given wholly distinct recognition.

7.3 FISHING AND HUNTING QUOTAS

Transferable development permits and water rights will have an indirect effect on the state of biodiversity. A much more direct and well-known mechanism is hunting or fishing quotas. Like the permits discussed above, they rely on there being an effective background prohibition on an activity, in this case the exploitation of a natural resource by hunting or fishing, against which a limited right to operate can be authorised. By restricting the number of animals, birds or fish that can be taken, the

---

[13] The existence, nature and extent of the right to water has itself been the subject of much debate; see Salman Salman and Siobhán McInerney-Lankford, *The Human Right to Water* (World Bank, 2004); Catarina de Albuquerque, 'Water and sanitation are human rights: why does it matter?' in Laurence Boisson de Chazournes, Christina Leb and Mara Tignino (eds), *International Law and Freshwater: The Multiple Challenges* (Edward Elgar Publishing, 2013).

[14] Godden, 'Water law reform in Australia and South Africa' (n. 11) 191–2; Water Act 2007 (Cth).

[15] Water Amendment (VEWH) Act 2010 (Vic.) amending Water Act 1989 (Vic.).

[16] Erin O'Donnell, 'Institutional reform in environmental water management: the new Victorian Environmental Water Holder' (2011) 22 *Water Law* 73.

impact on biodiversity is restricted, and since there are always risks of by-catch and disturbance to other species, any restriction on hunting activity and fishing is likely to have additional benefits. At the same time, the transferability of the quota adds greater flexibility than if the rights were tied to individuals, whether in their capacity as landowner or independently. Moreover, although the holders of the rights are not directly involved in the management of the ecosystem involved, they do have an interest in maintaining or enhancing its health, a legally recognised interest which means that they can take action to protect it, since this is essential if the resource that they wish to exploit is going to survive.

The fact that hunting and fishing rights could be severed from the land where the rights could be exercised was an early form of recognition for privately held rights in relation to biodiversity, distinct from those attached to landownership. For example, in Scotland[17] the right to all salmon fishing was vested in the Crown but could be granted quite independently of any other rights over land and water, a position that has survived the far-reaching reforms of land law at the start of this century.[18] In modern times, the most active discussions have tended to be in relation to marine fishing quotas, where the extent and nature of the rights have been explored.[19] Various schemes around the world have granted fishermen individual transferable quotas, entitling them to a share of the allowable catch in a particular area.[20] A recent review by Arnason supports the potential of strong individual rights to achieve economic efficiency in the exploitation of fishing resources, but finds that this approach is 'not sufficient for setting the socially optimal [Total Allowable Catch], ensuring the optimal use of the ecosystem, or harmonizing fishing with conflicting uses of marine resources'.[21]

---

[17] Except in Orkney and Shetland where udal law (the legacy of the Viking settlement of those areas) provided differently.

[18] Scottish Law Commission, *Report on Abolition of the Feudal System*, No. 168 (1999) paras 7.14–7.18.

[19] *United Kingdom Association of Fish Producer Organisations* v. *Secretary of State for Environment, Food and Rural Affairs* [2013] EWHC 1959 (Admin); see Chapter 2, section 2.9.

[20] E.g. the case studies reported in FAO, *Case Studies on the Allocation of Transferable Quota Rights in Fisheries*, Fisheries Technical Paper 411 (FAO, 2001).

[21] Ragnar Arnason, 'Property rights in fisheries: how much can individual transferable quotas accomplish?' (2012) 6 *Review of Environmental Economics and Policy* 217.

Such a conclusion points to a useful but necessarily limited role for such quotas. A quota system must operate within a wider framework where the major decisions on what is to be controlled and the level of permitted exploitation are reached on a different basis, and inevitably creates problems over coherence, since by design a system of transferable quotas cannot dictate exactly how activity is to be spread across the relevant field. As with so many of the mechanisms discussed here, they may have a part to play in conserving biodiversity, but only as one element in a wider framework.

7.4 IMPACT FEES

A quite different way of trying to ensure that something is provided in return for the loss to biodiversity arising from development is the levying of impact fees. These are one-off fees collected at the time a development is approved, conventionally representing a contribution to the cost of the infrastructure needs created by the new development. This approach could be extended so that such a fee is levied on the basis of the impact development is having on biodiversity, with the fees raised being dedicated not to infrastructure provision but to the conservation and enhancement of biodiversity.[22] Such a scheme shares some elements with both taxation and offsetting schemes, yet has more confined ambitions. A series of questions have to be answered in the design of impact fees, which help to place them in the wider catalogue of responses to the effects of development, and the evolution of the community infrastructure levy in England can illustrate many of these.

A first feature is that impact fees tend to emerge from the land use planning system. Although financial levies are involved, they are not introduced as part of the tax system, but are an offshoot of planning law.[23] In the English context, the starting point was the ability to grant permission for development subject to conditions, but with limitations on

[22] The US wetland mitigation scheme includes an 'in-lieu fee program' that in some circumstances allows fees to be paid to authorised projects in place of direct offsetting or use of a mitigation bank: *Federal Guidance on the Use of In-Lieu-Fee Arrangements for Compensatory Mitigation under Section 404 of the Clean Water Act and Section 10 of the Rivers and Harbors Act* (2000), available at www.fws.gov/habitatconservation/Corps%20In-lieu-fee%20guidance.pdf.

[23] In the US context, where the legality of various forms of 'exactions' has been the source of much litigation, this is expressed as coming under the 'police power'; *Berman* v. *Parker*, 348 US 26 (1954).

what those conditions could require.[24] In particular, conditions must focus on the development site itself and cannot call for the developer to make contributions to the wider infrastructure beyond the development site (even though the need for this might be provoked by the development).[25] However, the law also allows developers to bind themselves to the planning authority by means of a legal obligation which must still have a planning purpose related to the development in question, but can extend beyond what conditions can cover, for example, including contributions to roads in the neighbourhood to cope with the additional traffic generated by a development.[26] The source of the legal authority for such obligations and contributions, resting on the planning laws, ensures that there remains a connection between the development and any fee or other contribution obtained, distinct from general taxation. Views on such contributions differ, echoing the wider discussions about many of the mechanisms discussed in this book. One view is that they inappropriately allow developers to buy permission by offering incentives to (always) cash-strapped local authorities, whereas others see this as a welcome way of ensuring that the 'polluter pays', with developers bearing at least part of the burden that their (usually very profitable) activities are imposing on the community.[27]

Once the notion is accepted of developers contributing to the wider impact of their development, the question arises of the quantification of this. Given the emergence of these contributions from discussions about meeting the needs created by individual planning applications, it was inevitable that these were negotiated individually. This allows the obligations undertaken by the developer to be tailored specifically to the project in mind but can be criticised for uncertainty and unfairness between different developers and projects, as well as creating difficulty when the impact to be tackled is a cumulative one, with each individual proposal not of major significance in itself. Moreover, transaction costs are

---

[24] *Newbury District Court* v. *Secretary of State for the Environment* [1981] AC 578 (HL).

[25] *Hall & Co. Ltd* v. *Shoreham by Sea Urban District Court* [1964] 1 WLR 240 (CA).

[26] Town and Country Planning Act 1990, ss. 106–106C; *Tesco Stores Ltd* v. *Secretary of State for the Environment* [1995] 1 WLR 759 (HL). Contributions with a biodiversity aspect may be included, as shown by the rival offers to provide a bird-watching hide and wildlife habitat in *R* v. *Plymouth City Council, ex parte Plymouth and South Devon Cooperative Society* (1994) 67 P & CR 78 (CA).

[27] This has been a long-running debate; see e.g., Martin Loughlin, 'Planning gain: law, policy and practice (1981) 1 *Oxford Journal of Legal Studies* 61.

inevitably higher in the case of individually negotiated agreements. This has led to the introduction of the community infrastructure levy in England and Wales, under which authorities establish a charging schedule as the basis on which a standard-rate contribution from developers is to be collected.[28]

The making of the charging schedule is subject to considerable procedural formalities to ensure openness and stakeholder involvement.[29] Moreover, the basis on which the charge is set is limited. The authority must strike a balance between the likely cost of the infrastructure required to support the development of its area, taking into account other sources of funding, and the potential effects of levying the charge on the economic viability of development across its area.[30] Such provisions help to characterise the charge as something distinct from the general taxation system, where the rates are set by reference to wider criteria than those directly linked to the taxable activity itself (for example, the level of taxes on cigarettes in part reflects the costs imposed on the health system but are not calculated on the basis of those costs). In this way the community infrastructure levy can be seen as more like an offsetting scheme, endeavouring to balance the costs imposed by the permitted activity, but looking at the development of the area as a whole, rather than the consequences of each particular project in isolation.

The introduction of a standard charging scheme is intended to move away from the uncertainty, inconsistency and transaction costs of individually negotiated agreements, but the rate is not wholly uniform. The rate can vary according to the location of the development, its intended use, the internal area of the development (as opposed to its total 'footprint') and the intensity of the development.[31] Such scope for variation still means, of course, that the detailed impact of the specific development is not considered, moving the levy away from an offsetting model to one of taxation. The criteria for such variation might require careful consideration if a scheme were applied to collect a contribution based on the impacts on biodiversity rather than demands on infrastructure. Nevertheless, the fundamental provision is for the charging of a set

---

28 Planning Act 2008, Pt 11; Community Infrastructure Levy Regulations 2010, SI 2010/948.

29 Planning Act 2008, ss. 212–14; Community Infrastructure Levy Regulations 2010, regs 11–30.

30 Planning Act 2008, s. 211; Community Infrastructure Levy Regulations 2010, reg. 14.

31 *Ibid.* reg. 13.

rate per square metre of the development,[32] as opposed to making any more refined calculation of the precise impact of the project itself.

Once a charge has been levied, the question is what it can be used for. This is a major element in the nature of the scheme since the restricted uses to which the charge can be put is one of the major features which again marks it as something different from taxation (in a fiscal system where hypothecated taxes are not generally accepted). The charge must be used to 'support ... development by funding the provision, improvement, replacement, operation or maintenance of infrastructure', including roads and transport facilities, flood defences, schools and other educational facilities, medical facilities, sporting and recreational facilities and open spaces.[33] Again, if a biodiversity impact fee were to be introduced, a catalogue of biodiversity-related purposes for which the funds could be spent would have to be established.

Impact fees thus lie somewhere between offsetting schemes and taxation approaches to biodiversity. They operate purely on the basis of financial contributions and expenditure yet their design calls for many of the same issues to be addressed as are required for offsetting, notably identifying the activities that are to be caught by the scheme, the basis on which any contribution is to be calculated, and what counts as a legitimate use of the proceeds. Where impact fees may score over either offsetting or taxation is that the basic mechanism may already be in place so that with little or no adjustment the impacts on biodiversity can be taken into account and be addressed in this way. As an example, Scottish Borders Council has been effective in using existing mechanisms to gather a contribution from wind-farm developers in its area and applying this to further a number of large-scale conservation projects, offsetting the impact of the wind-farms on wildlife at the previously undeveloped sites affected.[34] With a clear and published policy on such matters, a framework was established which enabled all parties to know what was expected and to simplify the discussion and negotiations, whilst the existence of suitable long-term conservation projects in the area provided appropriate means of spending the contributions gathered. In doing so, it was possible to rely on the reputation and integrity of the partners

---

32 *Ibid.* reg. 12.

33 Planning Act 2008, s. 216.

34 A convenient summary is provided in their successful application for the Scottish Awards for Quality in Planning (2012), see www.scotland.gov.uk/Resource/0042/00426294.pdf. The authors are grateful to Dr Andy Tharme and his colleagues in Scottish Borders Council for the opportunity to discuss with them their policy, practice and experience.

involved to provide reassurance as to the future provision of biodiversity benefits rather than seeking to create specific legal guarantees of the sort that a stricter offsetting approach might demand.[35] In this way the practical results of an offsetting scheme were achieved through a much simpler impact fee mechanism.

[35] See Chapter 4, section 4.5.2.

8. Ethical issues

The previous chapters of this book have focused on the practicalities of new mechanisms for biodiversity conservation. These approaches apply market-based ideas to our dealings with nature and often involve human intervention to create or enhance habitat (in return for payment or to provide an offset). The discussion so far has hinted at, but not examined closely, some of the ethical questions which affect any consideration of such approaches and their acceptability.[1] A thorough examination of these issues is beyond the scope of a single chapter in any book, but what follows endeavours to give a fuller flavour of some of the issues which will colour responses to the mechanisms discussed here and of whether and how far they are viewed as an appropriate response to the ecological crisis we face. There are several dimensions to be considered. Some relate to the place of nature and biodiversity in our fundamental world view and challenge the extent to which nature should ever be treated on the basis of its value to humans. It is also noted that all human intervention, even with the best of intentions, entails a loss to what is currently in place, which itself has value, whether because of its naturalness or because the results of past human intervention themselves embody a rich social and cultural heritage. Others focus more narrowly on the use of market mechanisms, concerning both the fundamental appropriateness of this approach and, even where some scope for it is accepted, concern over how its use can itself affect the way things are perceived. A brief chapter such as this can only hope to scratch the surface and pick out a few examples of a broad and rich vein of critical thinking. What is vital, though, is that these issues are recognised and

[1] Although the consultation responses on biodiversity offsetting proposals in England concentrated on specific issues, analysis has revealed that contrasting views are often built on fundamentally different value frameworks; S. Sullivan and M. Hannis, 'Nets and frames, losses and gains: value struggles in engagements with biodiversity offsetting policy in England' (2015) 15 *Ecosystem Services* 162.

taken into account as policy develops. The choice to adopt new mechanisms, or to continue with more established approaches, which themselves embody choices with ethical dimensions, should not be viewed as one of technical effectiveness and efficiency alone.

8.1 LAND ETHIC, DEEP ECOLOGY AND WILD LAW

There are many reasons why humans do and should care about the environment and the living beings and communities with which we share the planet.[2] Many of these are based on the benefits that accrue to humans from nurturing a healthy environment: the analysis of ecosystem services is, after all, about considering the services provided to humans. Such justifications would tend to be supportive of measures that allow biodiversity to be managed, conserved or enhanced, even though this involves a substantial degree of human intervention, conscious habitat management and even substitution. There are, though, other views which would reject outright the notion that the natural world is amenable to being shaped by humankind for its own benefit.[3] Such views would see much of our existing conservation effort, let alone the greater adoption of market mechanisms such as biodiversity offsetting, as fundamentally misconceived, since it still rests on an anthropocentric view[4] valuing only the benefits (tangible and intangible)[5] that humans can receive from biodiversity.[6] Such an approach can be seen as offending against what should be at the very heart of our relationship with the natural world.

---

[2] For a convenient overview see Alexander Gillespie, *International Environmental Law Policy and Ethics* (2nd edn, OUP, 2014).

[3] For the sake of argument and brevity, views which criticise the current dominant world-view are presented in a simple form, without subjecting them in turn to thorough critique.

[4] For present purposes, the division between 'anthropocentric' and 'eco-centric' views is used merely as a convenient shorthand to capture a broad distinction, without analysis of the many issues these terms throw up, such as how far any human thought can escape an anthropocentric bias; Eric Katz, 'Against the inevitability of anthropocentrism' in Eric Katz, Andrew Light and David Rothenberg (eds), *Beneath the Surface: Critical Essays in the Philosophy of Deep Ecology* (MIT Press, 2000).

[5] Claude Gascon and others, 'The importance and benefit of species' (2015) 25 *Current Biology Review* 431.

[6] Matthias Schröter and others, 'Ecosystem services as a contested concept: a synthesis of critique and counter-arguments' (2014) 7 *Conservation Letters* 514.

There are many strands within the views that would call for a reassessment of that relationship and just a flavour of a few of these can be given.

In developing his Land Ethic, Leopold argues that: 'We abuse land because we regard it as a commodity belonging to us' rather than as 'a community to which we belong'.[7] Man may be 'the captain of the adventuring ship' but is 'hardly the sole object of its quest'[8] and must change from being 'man the conqueror' to 'man the biotic citizen',[9] from 'conqueror of the land-community to plain member and citizen of it'.[10] In the same way as slaves moved from being viewed as property to being part of our ethical community, the wider biotic community must come within our ethical compass.[11] Once this view is taken, thinking about what is 'ethically and esthetically [*sic*] right, as well as what is economically expedient'[12] would point towards a rejection of any conservation measures based purely on market-led approaches. Mechanisms such as offsetting which accept losses of biodiversity (albeit with compensating gains) offend against the fundamental tenet of the Land Ethic that: 'A thing is right when it tends to preserve the integrity, stability, and beauty of the biotic community. It is wrong when it tends otherwise'.[13] Preserving the integrity of the biodiversity that exists is what is demanded, not treating it as something to be traded, substituted and replaced.

A reconceptualisation of the relationship between humans and nature is a feature of deep ecology. Deep ecology can be seen as a fairly broad movement or platform rather than a specific set of ethical tenets[14] but central to this approach is a view of nature very different from one which would accept it as the subject of any form of trading or comparative valuation. Key ideas are stated in the first elements of the 'platform for the deep ecology movement' set out by Naess:

---

[7] Aldo Leopold, *A Sand County Almanac and Sketches Here and There* (OUP, 1949) viii.

[8] *Ibid.* 110.

[9] *Ibid.* 223.

[10] *Ibid.* 204.

[11] *Ibid* 201–4.

[12] *Ibid.* 224.

[13] *Ibid.* 224.

[14] 'There can be many different ecosophies, many different ultimate world views regarding human activity and life in the environment … one reason for the existence of the platform of the deep ecology movement is to provide common ground in the realm of practical activity for a wide variety of personal philosophical positions'; Katz, Light and Rothenberg, *Beneath the Surface* (n. 4) xii.

(1) The flourishing of human and non-human life on Earth has intrinsic value. The value of non-human life forms is independent of the usefulness these may have for narrow human purposes.
(2) Richness and diversity of life forms are values in themselves and contribute to the flourishing of human and non-human life on Earth.
(3) Humans have no right to reduce this richness and diversity except to satisfy vital needs.[15]

There are various 'depths' of ecology and not everyone who has sympathy with these initial elements may be equally committed to the following ones, which support a decrease in the human population to allow non-human life to flourish and proclaim an obligation on supporters to participate in attempting to bring about the necessary changes to implement all the elements identified. Nevertheless, however far one is prepared to follow these ideas, they present a perspective which calls for a fundamental reduction of the human impact on the natural world. Conservation measures that are simply about accommodating biodiversity within an environment shaped predominantly to meet human desires would fall far short of what is required, and trading and offsetting nature would be anathema.

Wild Law and Earth Jurisprudence present further analyses which embrace an ecocentric view.[16] This viewpoint again sees all elements of the Earth Community, not just humans, as having rights which must be respected:[17]

1. The natural world on the planet Earth has rights, which come with existence. These rights come from the same source from which humans receive their rights, from the universe that brought them into being.
2. Every component of the Earth community has three rights: the right to be, the right to habitat, and the right to fulfil its role in the ever-renewing processes of the Earth community.[18]

---

[15] Arne Naess, *Ecology, Community and Lifestyle* (David Rothenberg (trans), CUP, 1989) 29.

[16] Cormac Cullinan, *Wild Law: A Manifesto for Earth Justice* (2nd edn, Green Books, 2011).

[17] For a brief exploration of arguments furthering the rights of animals and nature, as opposed to merely humans, see Catherine Redgwell, 'Life, the universe and everything: a critique of anthropocentric rights' in Alan Boyle and Michael Anderson (eds), *Human Rights Approaches to Environmental Protection* (OUP, 1996).

[18] Thomas Berry, 'Rights of the Earth: we need a new legal framework which recognises the rights of *all* living beings' in Peter Burdon (ed.), *Exploring Wild Law: The Philosophy of Earth Jurisprudence* (Wakefield Press, 2011) 229.

Accordingly:

> [N]ature deserves to be valued for itself, because it is and exists, not because of the value we place on it as a resource. Humans are not granted better or more rights because they contribute more or less to society. They are granted rights in recognition of their existence, equality and for their protection.[19]

Again, therefore this requires that our dealings with biodiversity should be on a more equal footing, respecting nature rather than seeing it as something which is subject to human control.

The Wild Law approach seems to accept a more gradualist approach to the transformation of our thinking, laws and governance. It welcomes elements in the existing laws which show aspects of a Wild Law approach and discusses ways of going about 'the challenge of re-conceptualising the nature and purpose of government'.[20] Nevertheless, existing conservation measures and the options for change discussed here are not supported as the way forward: 'I do not have faith in ecological modernisation, markets and sustainability assessments as the means to achieve a viable future because they are all based on what I regard as fundamentally flawed premises and a philosophy of exploitation.'[21] In particular, new developments would be rejected when they appear to continue a commodification of nature, rather than contributing to the change in mind-set needed to reflect the place of humans as members of the Earth community: 'biodiversity offsetting fosters the illusion that we can buy and sell that which we do not truly own: the wild'.[22] From this perspective, therefore, it is a more radical transformation of the laws governing our dealings with nature which is required, not an extension of legal devices still based on a flawed relationship.

What this very brief sketch shows is the presence of ethical viewpoints which call for our relationship with biodiversity to begin from a new starting point. From those perspectives, humans are part of the natural world, not its masters, and any mechanisms which suggest that nature can be bought and sold or valued merely in terms of its contribution to humankind are fundamentally misconceived. In particular, biodiversity

---

19 Begonia Filgueira and Ian Mason, 'Wild law: is there any evidence of Earth jurisprudence in existing law?' in Burdon, *Exploring Wild Law* (n. 18) 195.

20 Cormac, *Wild Law* (n. 16) ch. 14.

21 Cormac Cullinan, 'Finding our way to a viable future: a response to Professors Warren and Lee' (2006) 18 *Environmental Law and Management* 18, 23.

22 Nicole Rogers, 'Where the wild things are: finding the wild in law' in Burdon, *Exploring Wild Law* (n. 18) 185.

offsetting would be rejected as being based on a view that humans are entitled to pick and choose what elements of nature should survive (already a feature of conventional approaches which identify the particular sites and species most worthy of protection) and to treat local habitats and ecological communities as commodities to be traded. Schemes based on payment for ecosystem services can be similarly criticised as seeing nature purely in terms of what it can do for humans. The reduction of nature to financial terms could be seen as representing a capitulation to the economic, human-centred view of the world which has led to the degradation of our environment. A more positive view might also be proposed, seeing such measures as a belated recognition that there is some value in nature, and as a first (and by itself very inadequate) step towards the integration of biodiversity into our ethical system.

8.2 (RE-)CREATING THE NATURAL?

A narrower way of approaching these issues is to concentrate on the values involved when humans intervene in the natural world, as required for the habitat (re-)creation which is a feature of many offsetting or PES projects. The essence of the argument is summed up in the title of Elliot's article 'Faking Nature'.[23] Here, he argues that even if it were possible effectively and completely to restore a habitat which had been destroyed,[24] something would still have been lost even though the components and functionality of the habitat matched that of its earlier state. The new habitat would not be exactly the same as what has been lost, and the mismatch would be of significance; there is a difference between a genuinely natural habitat and what is in effect an artefact, the product of human action, not of natural processes. In the same way as an identical replica does not have the intrinsic values of an original Old Master painting, so a restored or recreated habitat does not have the same qualities as one which has not been shaped by human interference. It is not just the final product that counts; 'the causal genesis of forests, rivers,

---

[23] Robert Elliot, 'Faking nature' (1982) 25 *Inquiry* 81. The idea is developed in his book of the same title (Routledge, 1997). See also Eric Katz, 'Further adventures in the case against restoration' (2012) 34 *Environmental Ethics* 67.

[24] This is a 'very big if' given the vast number of features, components and processes which would have to be matched to produce an identical replica, and the difficulty of achieving even substantial success; see Chapter 2, section 2.1 and Chapter 4, section 4.2.

lakes, and so on is important in establishing their value'.[25] The consequence is that even our efforts to restore and respect nature, giving it priority over conflicting human desires, represent 'a continuation of the paradigm of human scientific and technological mastery over natural processes'.[26]

Deliège picks up this argument in the context of a dispute over whether it is acceptable deliberately to plant specimens of missing flora in the restoration of heathland as opposed to waiting for natural re-colonisation to take place.[27] There are contrasting views on what precise goal is being sought in the quest to end up with a healthy area of heathland; a 'natural' heath which has not been interfered with or one which has achieved the fullness of biota attainable in the conditions, even though this calls for some 'gardening' to reach that state? Again, authenticity can become a key issue, with value seen to be residing in the characteristic of the habitat as a relic from the past, so that a replacement will never have the same worth: 'A heath garden might simulate an authentic heath, but it is not one, and it never can replace the real deal'.[28] An added complication here is that the sort of heath in question does not represent the natural apex vegetation for the region but is the result of the ancient clearance of woodland and of continuing management to prevent scrub and woodland taking over; the greater biodiversity achieved through the presence of the heath is probably not what would result if nature were completely left to itself.

A separate strand queries projects to (re-)create natural environments from a more human perspective. This can be illustrated by Drenthen's account of reaction to plans for managing watercourses in the Netherlands. Policy switched from favouring large-scale 'hard engineering' projects to working with, rather than against, the elements, allowing certain areas to be inundated and develop as wetland reserves as the means of diminishing the harmful impacts of floods.[29] Both options, though, attracted opposition on the basis that they would eradicate the existing man-made landscape, the product of centuries of human endeavour in managing water resources on a local scale. Both approaches were equally destructive of the history of the land, the 'traditional' landscape

---

25 Elliot, 'Faking nature' (n. 23) 89.

26 Katz, 'Further adventures' (n. 23) 72.

27 Glenn Deliège, 'The cinquefoil controversy: restoring relics between managers and purists' (2010) 32 *Environmental Ethics* 17.

28 *Ibid.* 32.

29 Martin Drenthen, 'Ecological restoration and place attachment: emplacing non-places?' (2009) 18(3) *Environmental Values* 285.

that embodied the vision and effort of the ancestors of today's local residents and which created the sense of place felt by those living in the area.

This shows that any major intervention in how the land is shaped, used and managed can obliterate the cultural as well as the natural heritage and whilst the 'rewilding' of an area may be seen as restoration of its natural value, at the same time it can destroy much of its historical and social value. Again, the question is raised of what it is we value about particular places and why, and the balance between natural and cultural significance. In many places it has taken centuries of hard labour to shape the natural environment so that it can provide for those living in the area, and there can be a real sense of loss, and indeed of a disrespect to past generations, if the land is abandoned to nature once more in order to achieve the objectives of an offsetting or PES scheme promoting biodiversity. We can seek a sense of place through attachment to the human history of the land or to its biological qualities or through the appreciation of a 'palimpsest landscape'[30] with multi-layered traces of different elements and values. The simple point, perhaps, is to remember that even taking commercial and practical considerations out of account, not everyone will see the 'rewilding' of land as a desirable objective.

8.3 LIMITS OF THE MARKET

Market-based approaches can be presented as representing a world-view that is the opposite of what has been discussed above. The conversion of biodiversity into commodities[31] which can be used by humankind as the basis of payments and trades seems the antithesis of a view which values all members of the Earth community, as the apotheosis of the anthropocentric view of the world that surrounds us. Such stark condemnation would be unfair, since there are many variations within the range of views which can envisage some role for an economic approach within environmental law. Nevertheless, introducing markets in any form alters not just the policy tools that are deployed but deeper relationships. Steps in that direction are 'not first of all about moving from public policies to

[30] *Ibid.*

[31] The valuation of ecosystems and ecosystem services does not by itself necessarily entail treating them as commodities; Robert Constanza and others, 'Changes in the global value of ecosystem services' (2014) 26 *Global Environmental Change* 152, 157.

market allocations. It is more about a reconfiguration of state-market-community relationships'.[32] Without going into a full consideration of the justification for and criticism of a market-based approach,[33] a few points of specific application to biodiversity can be made as further factors to weigh in determining the appropriateness of utilising mechanisms with a market element as part of our response to the need for the conservation of biodiversity.

Concerns affect both the applicability and the operation of markets, and although some of the latter may initially seem to be essentially practical problems, they often suggest a mismatch at a higher level. Thus, Robertson's account of the difficulty of identifying plants which have importance as indicator species for the purposes of the US wetland mitigation scheme rapidly turns to consider the limits of ecological certainty in contrast to the precision required for commodity trading:[34]

> The difference between selling ecosystem services and selling loaves of bread is that legal and capital logics require information about ecosystem services that scientists *cannot provide* in a uncontroversial way ... Whereas the metrical evaluations involved in standard commodities are the product of a successful and stable articulation of science with capital, the metrical evaluation of wetland services is the product of a constantly shifting and unstable articulation, producing information that works within capital only by virtue of a set of ad hoc logics employed by those who populate the forums of articulation.[35]

Even if improvements can be made in the tailoring of the currencies of trade to account for such difficulties, such steps will inevitably increase transaction costs, counteracting the efficiency which is one of the main arguments for adopting a trading approach in the first place.[36] The difficulty, however, is not just one of inexperience or lack of refined

---

[32] Arild Vatn, 'An institutional analysis of payments for environmental services' (2010) 69 *Ecological Economics* 1245, 1252.

[33] For a brief summary see LeeAnne Edmonds Applegate, 'The suitability of government for economic valuation of natural resources and environmental harm' (2006–07) 21 *Journal of Natural Resources and Environmental Law* 29.

[34] Morgan Robertson, 'The nature that capital can see: science, state and market in the commodification of ecosystem services' (2006) 24 *Environment and Planning D: Society and Science* 367.

[35] *Ibid.* 382.

[36] James Salzman and J.B. Ruhl, 'Currencies and commodification of environmental law' (2000–01) 53 *Stanford Law Review* 607, 636.

techniques, but an inherent mismatch between the sort of answers that ecology can provide and the sort of information that markets need.[37]

Such a mismatch is also identified by Salzman and Ruhl, who note that whereas standard commodity markets, and successful environmental trading schemes, operate on the basis of fungible goods, expansion into other areas of environmental trading faces an obstacle in the non-fungibility of the objects being dealt with.[38] This non-fungibility can be of several sorts: of space (in so many ways a hectare of land in one place is not the same as one in another location); of type (wetlands are not the same as sand dunes); and of time (the features of a mature forest felled today are not immediately replaced by new planting). In relation to biodiversity these non-fungibilities are particularly marked, but they are present even in schemes for atmospheric emissions where although the focus may be on the global concentration of gases, the balance between diffuse and concentrated sources and between the characteristics of the various gases converted into carbon dioxide equivalents do still make a difference in the real world. The authors do not reject a role for markets but conclude that:

> The commodity model of trading cannot, we believe, sufficiently satisfy the demands posed by the trading of nonfungible environmental amenities … Exchanging environmental apples for oranges may be a beneficial policy choice, but let us be honest about what [environmental trading markets] are trading.[39]

Other analyses of market-based approaches point to circumstances where they can have adverse consequences and so should not be applied. Satz notes that:

> markets can have not only economic effects but also political and cultural effects. Particular markets can crowd out altruism, enable hierarchical relationships between market agents, and even undermine the conditions for a democratic society.[40]

---

37 Nicolás Kosoy and Esteve Corbera, 'Payments for ecosystem services as commodity fetishism' (2010) 69 *Ecological Economics* 1228.

38 Salzman and Ruhl, 'Currencies and commodification of environmental law' (n. 36).

39 *Ibid*. 693–4.

40 Debra Satz, *Why Some Things Should Not be for Sale: The Moral Limits of Markets* (OUP, 2010) 208.

In an environmental context,[41] her specific 'social democratic thesis: that certain goods need to be provided outside the market if citizens are to be equals'[42] might provoke an exploration of links with ideas on environmental justice.[43] Yet she also makes the wider point that the very fact of subjecting something to market forces can transform our relationship with it. This point is also made by Sandel,[44] who notes that the extension of market thinking into an increasing range of situations in itself changes norms and attitudes,[45] and that this inevitably raises moral concerns.

The changes brought about by introducing market-based thinking can be illustrated in situations where there are harmful consequences for others for deviating from a norm of behaviour. If payments are introduced as a way of redressing the balance for any deviation from the norm, this can be viewed either as a fine for transgression, confirming the 'wrongness' of the conduct, or as a fee permitting an exception from the norm, altering the attitude to the norm itself.[46] Causing the adverse consequences is no longer a wrong but a choice, open to those with the willingness, and importantly the ability, to pay. Where there is broad recognition that society has agreed that the conduct is undesirable, any fine will tend be seen as a penalty. Some motorists may not be supportive of speed limits and be willing to disregard them, but there is no suggestion that they think along the lines that a willingness to pay the fine if caught creates a personal entitlement to break the speed limit.[47] Where the 'wrongness' of the conduct is less agreed, there is a greater

---

[41] Much of the argument, somewhat to her surprise, focuses on 'markets … closely connected to the human body'; *ibid.* 210.

[42] *Ibid.* 208.

[43] David Schlosberg, *Defining Environmental Justice: Theories, Movements, and Nature* (OUP, 2007).

[44] Michael J. Sandel, *What Money Can't Buy: The Moral Limits of Markets* (Allen Lane, 2012).

[45] '[M]arkets change the character of the goods and social practices they govern'; *ibid.* 120.

[46] The example given is of collecting children from nursery care later than the scheduled time, where the introduction of a 'late fee' actually increased the number of late collections as parents saw this as an extra charge for an extra service that was being provided, not as a penalty for an unfair imposition on the staff, an attitude to lateness that persisted even after the payments were removed; *ibid.* 119, citing U. Gneezy and A. Rustichini, 'A fine is a price' (2000) 29 *Journal of Legal Studies* 1.

[47] Charles Musselwhite and others, *Understanding Public Attitudes to Road-User Safety, Literature Review*, Road Safety Research Report No. 112 (Department for Transport, 2010).

likelihood that any payments will be viewed as the price for doing something, rather than the penalty to be paid for doing wrong.

Thus, in the same way as taxes on emissions can be seen as allowing people to buy a 'right to pollute', biodiversity offsetting can be seen as allowing developers to buy a 'licence to trash' our natural heritage. Especially in view of the debates over where the appropriate balance lies between the environmental, economic and social elements of sustainability,[48] there can be arguments over when activity with some adverse consequences for biodiversity is properly seen as wrongful at all.[49] The more general point, though, is that the market approach may claim to be morally neutral but is inevitably making implicit judgements that have a significant moral dimension:

> When market reasoning travels beyond the domain of material goods, it must 'traffic in morality', unless it wants blindly to maximise social utility without regard for the moral worth of the preferences it satisfies.[50]

As Sandel shows, questions over the 'moral limits of markets' arise from the use of a market approach in many areas of life,[51] and there are no easy answers. One area where conflicting views are being reflected in continuing policy debates is in relation to water resources where there is a tension between the view of water as common heritage and as a commercial resource. On the one hand, it is said that:

> [Water] is not a commercial product like any other but, rather, a heritage which must be protected, defended and treated as such.[52]

---

48 Duncan French, *International Law and Policy of Sustainable Development* (Manchester University Press, 2005) ch. 2.

49 Sandel's discussion of tradable emissions permits reveals the ambiguities here, since he accepts that emitting carbon is not in itself objectionable, whereas excessive emissions are, and argues that allowing large emitters to buy their way out of changing their wasteful habits reinforces bad attitudes, even though in atmospheric terms it is the total level of emissions, not who is producing them and why, that matters; Sandel, *What Money Can't Buy* (n. 44) 72–9.

50 *Ibid.* 89.

51 In the environmental field Sandel discusses tradable pollution permits, carbon offsets and payments to hunt animals that are endangered (*ibid.* 72–84), and other examples include priority queues at airports, incentives to lose weight and payments to donate blood and human organs.

52 Directive 2000/60/EC of the European Parliament and Council of 23 October 2000 establishing a framework for Community action in the field of water policy, [2000] OJ L327/1, Preamble, para. 1.

On the other hand, it is said:

> Past failure to recognize the economic value of water has led to wasteful and environmentally damaging uses of the resource. Managing water as an economic good is an important way of achieving efficient and equitable use, and of encouraging conservation and protection of water resources.[53]

Giving a commercial value to water enables it to be recognised in the economic calculations that dominate so much of the modern world's activities and therefore to be treated as a serious consideration attracting considerable weight. But does 'playing that game' entail the abandonment of another scale by which we can measure and reflect the things that are of value to us as humans? For the conservation of biodiversity, itself declared to be 'a common concern of humankind',[54] does embracing market mechanisms offer an effective way to fight the pressures that have led to such serious decline, or involve abandoning the very values that make the effort worthwhile?[55]

8.4 AFTERWORD

This chapter has given only a brief glimpse into fascinating and important debates about fundamental issues affecting human values and our relationship with nature and other humans. Further avenues not explored include those concerning the precise meaning of, distinctions between and consequences of various approaches which utilise the economic framing of the environment, monetisation and commodification.[56] In all of this it must be remembered that views differ between individuals and over time. In some areas we have moved away from seeing things in

---

[53] Dublin Statement on Water and Sustainable Development prepared by the International Conference on Water and the Environment (1992), available at www.gdrc.org/uem/water/dublin-statement.html.

[54] Convention on Biological Diversity, Rio de Janeiro, Brazil, 5 June 1992, in force 29 December 1993 (1760 UNTS 79), Preamble, para. 3. The concept of heritage also appears in the EU biodiversity texts: *European Commission* v. *Portugal* (C-191/05) [2006] ECR I-6853; Council Directive 92/43/EEC of 21 May 1992 on the conservation of natural habitats and of wild fauna and flora, [1992] OJ L206/7, Preamble, para. 4.

[55] Laurence H. Tribe, 'Ways not to think about plastic trees: new foundations for environmental law' (1973–74) 83 *Yale Law Journal* 1315.

[56] Erik Gómez-Baggethun and Manuel Ruiz-Pérez, 'Economic valuation and the commodification of ecosystem services' (2011) 35 *Progress in Physical Geography* 613.

purely economic terms, favouring human rights over an economic view of slavery. In other areas, economic and financial perspectives have been allowed to play an increasing role in determining policy.[57] There are no easy answers, but a need to appreciate that these different ways of viewing the world exist and lie unexpressed, and perhaps often not even recognised, below the surface of the debates on how we should deal with biodiversity. The choices between different mechanisms, innovative or conventional, are not value-free but practical governance does not demand that we adopt and follow to its limit any one particular viewpoint. Both individual humans and human societies are remarkably good at making a success of life even while holding incoherent and contradictory views on all sorts of things.

57 'The financial evaluation of a man's life introduced by the life insurance industry was initially rejected by many as a profanation which transformed the sacred event of death into a vulgar commodity'; Viviana A. Zelizer, 'Human values and the market: the case of life insurance and death in 19th century America' (1978) 84 *American Journal of Sociology* 591.

9. Reflections

At the heart of the topics covered in this book lies the way in which biodiversity is valued. The catch is that the word 'value' can be used in quite different ways. Looking at the *Oxford English Dictionary*, the meanings given include:

> Worth based on esteem; quality viewed in terms of importance, usefulness, desirability, etc.;
>
> The material or monetary worth of something; the amount at which something may be estimated in terms of a medium of exchange, as money or goods, or some other similar standard; and
>
> That amount of a commodity, medium of exchange, etc., which is considered to be an equivalent for something else.[1]

This raises the question of whether valuing biodiversity is a matter of morality or of money; each of these leads on to quite different legal responses.

That question has become more pointed as recent years have seen the emergence of more sophisticated approaches to viewing nature in terms of ecosystem services[2] and natural capital.[3] Such developments create new ways of thinking about the value of biodiversity and hence new opportunities for biodiversity law and regulation. Without going as far as to convert everything into monetary terms, we now have methods for assessing the many ways in which humans benefit from the natural world. Moreover, such methods enable us to recognise the extent to

---

1 *Oxford English Dictionary Online* (OUP), www.oed.com/view/Entry/221253?rskey=CzRbRF&result=1#eid.

2 As developed in the various Millennium Assessment Reports, see www.millenniumassessment.org/en/index.html; UK National Ecosystem Assessment, *The UK National Ecosystem Assessment: Synthesis of the Key Findings* (UNEP-WCMC, 2011) and further reports available at the website of the UK National Ecosystem Assessment, http://uknea.unep-wcmc.org.

3 E.g. in the United Kingdom the work of the Natural Capital Committee in England, www.naturalcapitalcommittee.org/ and the Scottish Forum on Natural Capital, http://naturalcapitalscotland.com/.

which specific actions are enhancing or diminishing ecosystem services and natural capital and thus to assess gains, losses and equivalences. This form of assessment in turn enables us to consider the harms being done and benefits delivered in a way that allows regulatory frameworks to take account of such contributions (positive and negative) and to attribute credit and responsibility accordingly.

It becomes possible to think about biodiversity in terms of the 'polluter pays' (such as requiring developers to offset the biodiversity losses they cause) or the 'beneficiary pays' (such as requiring payment for ecosystem services, rewarding those who provide them). Perspectives based on valuing nature also open the way to a practical discussion of the idea of 'no net loss' as a minimum objective if we are to stem the continuing loss of biodiversity across the globe. With such concepts in place, maintaining and enhancing biodiversity can become integrated with, rather than standing wholly outside, the commercial and economic networks which dominate our world. Supplying biodiversity benefits can be appreciated (and potentially rewarded) as a productive use of land, not just a residual by-product where no alternative, more commercial use is possible. In turn, this allows new regulatory approaches to be adopted, moving away from the 'command and control' viewpoint which has shaped so much conservation law.

Whether such developments are wholly to be welcomed is a different matter. As noted in the Preface, this would in many ways have been an easier book to write (and possibly more stimulating to read) if the authors held and presented strong beliefs in favour of or against the desirability of the widespread adoption of market-based and other 'privatised' approaches to biodiversity conservation. On the one hand, market-based and related instruments might be seen as the solution to all the relevant regulatory problems, delivering an effective and efficient way of both enhancing biodiversity and fulfilling other human objectives whilst escaping the heavy burden of direct regulation by public authorities. They also create the potential for new sources of funding to support conservation efforts. On the other hand, any mechanisms which can be seen as seeking to attribute quantifiable value to nature and to think in terms of 'net' losses and gains might be seen as fundamentally misguided, as a commodification of nature and a denial of the very qualities which make biodiversity worth caring about. The end result of our study, however, is that we remain in the middle ground, recognising the potential for new approaches to improve on the current position but at the same time conscious of the pitfalls if they are applied too widely or without sufficient care.

In this we are not alone. To take just one example from many, the pragmatic response from many points of view is clear from the report of a workshop in Edinburgh involving stakeholders from backgrounds in conservation, development, regulation and academia.[4] This reveals both a cautious acceptance of biodiversity offsetting as a means of improving on the present position where nature simply loses out when development takes place and also a concern that such offsetting could be damaging unless operated within a clear mitigation hierarchy (with offsetting as the last resort where harm cannot be avoided or lessened) and with absolute protection for the most important sites. The practical difficulty of delivering and valuing biodiversity gains was appreciated, but with an acknowledgment that since the metrics could never be perfect what should be sought was a system that was 'good enough'. Views differed on various aspects of what a suitable scheme might look like, but there was agreement on the importance of standardisation and clear guidance so that everyone can know where they stand. Similar sentiments have been expressed at many workshops, conferences and meetings we have attended.

The tensions are clear within this conditional acceptance of deploying a wider range of legal mechanisms to further conservation and are reflected throughout the more detailed discussions in the preceding chapters. Resolving these is difficult, but it is obvious that any developments will have to be tailored to the individual context, based not least on the general legal background and the role which any innovations are intended to play. There is a big difference between a scheme seeking to supplement a strong and far-reaching set of laws that provide effective conservation for a wide range of species and habitats and one which is intended to be the principal element of the conservation framework. This is definitely not an area for 'one size fits all'. A careful balance has to be struck between the competing considerations, such as between accurately reflecting the variety and complexity of the natural world and devising a scheme that is simple and easy to operate, and between seeking strong guarantees of future results and establishing a scheme which is not viewed as so onerous that it deters potential participants. Such challenges are common to all discussion of regulatory innovation, but there are two features here which make choosing the way forward particularly difficult.

---

[4] Susan Davies, *Biodiversity Offsetting and Habitat Banking for Scotland: Notes from the Workshop* (Edinburgh Centre for Carbon Innovation, 2013), available at http://edinburghcentre.org/files/documents/SD_Biodiversity_Offsetting_and_Habitat_Banking_for_Scotland_FINAL_REPORT.pdf.

The first is the combination of uncertainty and the irreplaceable nature of the species and habitats at stake. As noted above,[5] uncertainty exists on many points and creates risks of things not working out as foreseen, but if we get things wrong we risk losing forever elements of the biodiversity which creates the world we share. There is uncertainty over how any human intervention is going to affect biodiversity, and especially over how effective projects can be in enhancing, restoring and (re)-creating habitats and ecological communities. The results of our actions will take decades to emerge. Yet giving credit in some form for such actions is central to payment and offsetting schemes, relying on the fact that positive results are being achieved. There is uncertainty over the appropriateness and operation of various methods of attributing and quantifying value (financial or other) to biodiversity. There is also uncertainty over how people will react to new approaches to regulation, as shown by the way in which the well-intentioned measures to incentivise the production of bio-fuels led to the damaging conversion of virgin land to agricultural production.[6] In the face of uncertainty in environmental matters, one response can be to turn to the precautionary principle, but this can be used both ways. Does a precautionary approach call for eschewing new regulatory interventions whose results are uncertain and might actually have adverse consequences, or does it demand that in the face of the 'serious or irreversible damage'[7] currently being suffered we push ahead with what we hope, but cannot guarantee, will succeed?[8]

The second is the long time-scales involved at many stages. The length of time before we can ascertain how the natural environment is responding to our attempts at enhancing or restoring biodiversity has a significant knock-on effect that limits the opportunity to carry out pilot projects as we seek to develop new legal frameworks. It will be decades or even centuries before we can tell whether programmes started today have

---

[5] See Chapter 2, section 2.1.

[6] Alison Campbell and Nathalie Doswald, *The Impacts of Biofuel Production on Biodiversity: A Review of the Current Literature* (UNEP-WCMC, 2009).

[7] This phrase is taken from the statement of the precautionary principle in the Rio Declaration: 'Where there are threats of serious or irreversible damage, lack of full scientific certainty shall not be used as a reason for postponing cost-effective measures to prevent environmental degradation': Rio Declaration on Environment and Development, June 1992, Principle 7, UN Doc. A/CONF.151/26 (1992) vol. I, (1992) 31 ILM 874.

[8] The meaning and application of the precautionary principle is, of course, subject to much debate; see e.g. the brief bibliography in Stuart Bell, Donald McGillivray and Ole Pedersen, *Environmental Law* (8th edn, OUP, 2013) 84.

really achieved the ecological results we hope for and how any offsetting system has affected the 'net value' of biodiversity and the balance between different habitats and species. Moreover, the length of time before consequences start to become clear also limits the scope for adaptive regulation, fine-tuning the regulatory system as we learn from experience. By the time results are apparent we will have missed many opportunities for building on the positive, or limiting the adverse, outcomes of the schemes initially adopted. In an area where there is so much uncertainty, experimentation and learning as we go might be expected to form a major strand of the way forward, but here the time-scales may impose some limits on what can be done through such an approach.

A further aspect of the time-scales is the strain this puts on legal, financial and administrative structures. Very long-term commitments tend to be avoided by public bodies, commercial organisations and individuals, who are all very aware of how much the world and their capacities and priorities may change over time. There is also a desire to leave successors with freedom of action, rather than being bound by the 'dead hand' of the past. Yet it is long-term, and even permanent, obligations which are central to some of the approaches discussed here. Creating legal structures which are likely to be able to ensure that obligations are observed and enforced decades after those who initially created them are gone is a real challenge, especially when some financial responsibilities are involved. In seeking models to follow, attention naturally turns to property law, where there are well-established mechanisms for dealing with long-term relationships, but these inevitably 'bring their own baggage' and limit the ways in which relationships can be shaped.[9]

There are undoubtedly significant design challenges to be grappled with if instruments based on attributing value to biodiversity are going to be introduced in a way which is effective, attractive and workable. In contemplating these, it is crucially important, though, that we do not think that the more innovative approaches have a monopoly of difficulty and problems. The more conventional 'command and control' approach itself embodies many similar choices which themselves are far from uncontestable, and flow from a view of the relationship between humans and nature which itself is very different from that promoted by the Land

[9] This is especially so when more than one element come together, as with the impact of seeking charitable status for conservation easements in the United States; see Chapter 5.

Ethic or Wild Law.[10] Any identification of species or habitats for special protection involves an element of comparative valuation. The strength of the limitations on the landowners' rights to deal with their own land as they wish involves choices between pursuing the public good of conservation and preserving private rights. The statutory procedures reflect choices on the extent to which public participation and individual initiative are going to be supported.

Above all, the current laws are not proving to be effective, with biodiversity loss continuing around the world. Better results could undoubtedly be achieved if laws of the conventional sort were more widely applied and more strictly enforced. But is simply doing more of the same, a bit better, the best solution? Aside from the broader reassessment of regulatory approaches in many fields,[11] the dominant legal focus on specific designated sites is being seen as too narrow and rigid in an increasingly dynamic world. Both the ecosystems approach and appreciation of the need for action to be taken at a landscape-scale move the focus away from concern for a number of sharply bounded nature reserves (and similar designations of habitat) towards caring for the health of the wider environment.[12] The notion of adaptive management[13] and the need for species to be given the scope to relocate in response to a changing climate similarly point away from a static network of designated sites and from efforts based on preserving current ecological communities.

The 'command and control' approach is not so well-suited to promoting this newer vision of what might be called 'diffuse conservation', but this is exactly where the 'privatised' mechanisms discussed here can make a contribution. Positive outcomes for conservation would be the result not of centrally determined preferences, priorities and programmes, but of the cumulative impact of many individual decisions. These decisions would be influenced by a range of measures which lead to biodiversity being recognised as having value and being taken into account as an important consideration as decisions are made affecting many aspects of the environment. At present the designated sites may have strong protection, but beyond their boundaries biodiversity is too often simply ignored. The law can help to change this by adopting

---

10 See Chapter 8.

11 See Chapter 1, section 1.5.

12 See e.g., Scottish Government, *2020 Challenge for Scotland's Biodiversity* (2013) ch. 1.

13 Nick Salafsky, Richard Margoluis and Kent Redford, *Adaptive Management: A Tool for Conservation Practitioners* (World Wildlife Fund, 2001).

structures which reflect the value of biodiversity in different ways and which address this imbalance, ensuring that biodiversity is one of the factors being juggled as plans and projects are developed and implemented, as opposed to being at best an afterthought once decisions have been made.

Around the world there are many schemes that employ the mechanisms examined in the preceding chapters and, especially if their results are properly monitored,[14] it will be possible to identify more clearly the strengths and weaknesses of various approaches and the impact of detailed aspects of the design of specific schemes. No regulatory mechanism, new or existing, is without flaws and there are several dangers we must avoid. Refusing to contemplate any use of market-based and related instruments because they are not perfect is as much an error as looking at the less than complete success of the current laws, deciding that 'something must be done' and rushing blindly to adopt new mechanisms simply because they offer an alternative 'something'. Those at one end of the spectrum must be shown that giving all biodiversity a monetary value will not by itself open the door to the solution for all the regulatory problems. Those at the other end must be persuaded that placing a value (monetary or not) on biodiversity does not mean abandoning other views of biodiversity but may be just a tool for ensuring that it is not ignored. Indeed, it may even be a first step leading to acceptance of a renewed relationship with the natural world, where nature is seen as more than just a resource to be exploited.

If biodiversity is to flourish, and with it the human race, more must be done to protect and enhance nature and this will require legal intervention. Such intervention can come in many forms and it seems inevitable that the best way forward will involve a combination of measures – traditional 'command and control' as well as market-based and related instruments. The trick is to determine what the conventional approach can do well and where different mechanisms that make use of our increasing analysis and quantification of the value of biodiversity can produce real improvements on the current position. The devil is in the detail and further consideration must be given to how mechanisms can be designed to maximise the benefits and minimise the risks. A degree of experimentation seems appropriate, in order to determine how individual measures, and combinations, will work in specific circumstances, but always mindful of the uncertainties involved and the long time-scales

---

14 See e.g. the concerns over the true effectiveness of the US wetland mitigation scheme, discussed in Chapter 4.

before results, good or bad, may become clear. This book, we hope, has helped to lay out some of the choices before us and will assist in the necessary discussions of the practical, and ethical, merits of the different legal mechanisms which are available as the route towards an ecologically sustainable future is planned.

Index